RECENT DEVELOPMENTS ON POWER INVERTERS

Edited by **Ali Saghafinia**

Recent Developments on Power Inverters
http://dx.doi.org/10.5772/65189
Edited by Ali Saghafinia

Contributors

Pedro Melin, Carlos Baier, Jaime Rohten, Jose R. Espinoza C., Marian Gaiceanu, Branislav Dobrucky, Abdelhalim Sandali, Cheriti Ahmed, Rubén Peña, Javier Riedemann, Ramón Blasco-Gimenez, Natarajan Prabaharan, Kaliannan Palanisamy

Published by InTech
Janeza Trdine 9, 51000 Rijeka, Croatia

Publishing Process Manager Dajana Pemac
Technical Editor SPi Global
Cover InTech Design team

First published June, 2017
Printed in Croatia
Legal deposit, Croatia: National and University Library in Zagreb

Additional hard copies can be obtained from orders@intechopen.com

Recent Developments on Power Inverters, Edited by Ali Saghafinia
p. cm.
Print ISBN 978-953-51-3231-8
Online ISBN 978-953-51-3232-5

For EU product safety concerns:
IN TECH d.o.o., Prolaz Marije Krucifikse Kozulić 3, 51000 Rijeka, Croatia,
info@intechopen.com or visit our website at intechopen.com.

Meet the editor

Ali Saghafinia was born in Esfahan, Iran, in 1973. He received the BSc degree in electronic engineering from Najafabad Branch, Islamic Azad University, Iran, in 1995, and the MSc degree in electrical engineering from Isfahan University of Technology (IUT) in 2001. He was a lecturer at the Department of Electrical Engineering, Majlesi Branch, Islamic Azad University, Esfahan, Iran, during 2002–2008. His PhD degree was awarded in 2013 by the University of Malaya, Kuala Lumpur, Malaysia, in electrical engineering. Dr. Ali Saghafinia passed a postdoctoral research fellow at UM Power Energy Dedicated Advanced Centre (UMPEDAC), University of Malaya, in 2013–2014. Dr. Ali Saghafinia was an assistant professor at Majlesi Branch, Islamic Azad University, in 2014–2016. Currently, he works at the Department of Electrical and Computer Engineering, Isfahan University of Technology (IUT), as a postdoctoral research fellow. His research interests include power electronics, electric motor drives, machine design, fault detection, and industrial engineering. He possesses 15 years of teaching experience and has authored or coauthored over 50 books, book chapters, and papers in international journals and conferences.

Contents

Preface

Power inverters as the most suitable solution to provide a variable voltage/current with adjustable magnitude and frequency have been widely used in industry for several applications including AC motor drives, AC uninterruptible power supplies (UPS), active harmonic filter, induction heating, photovoltaic (PV) applications, etc., to be single phase or three phases in their output. Depending on the existence of voltage or current source as the source for DC link, there are two models of the inverter, which are known as voltage source inverter (VSI) and current source inverter (CSI), respectively.

Generally, six-step method and pulse-width modulation (PWM) techniques are two popular methods for the inverters. However, the existing lower order harmonics of the six-step voltage wave, which causes large distortions of the current wave and voltage control by the line-side rectifier, are limitations of this switching method. Therefore, techniques such as pulse-width modulation (PWM) techniques, the soft switching techniques, pulse density modulation (PDM), etc. with regard to parameters such as cost, efficiency, lower harmonic distortion, and transient state are the best options for applications in high-performance inverters. Moreover, to control output voltage and switching losses, reduce the size of the reactive components and switching stress, as well as optimize the harmonics due to the power electronic switching, the mentioned techniques alone or along with some changes in the structure like reduction of the switching device and the several kinds of the multilevel inverters as compared to conventional two-level inverters have been applied to reach the mentioned goals and have a suitable output.

This book develops and presents some methods and structures for improving the power inverters for different applications in single-phase or three-phase output to recover the aforementioned problems in recent years. The reduction of the switching devices and multilevel inverters as changing structure for the power inverters and PDM and PWM methods as changing control methods for the power inverter are studied in this book. Moreover, power inverters are developed to supply open-ended loads. Furthermore, the basic and advanced aspects of the electric drives that are control based are taught for induction motor (IM) based on power inverters suitable for both undergraduate and postgraduate levels.

The main objective of this book is to provide the necessary background to improve and implement the high-performance inverters. Once the material in this book has been mastered, the reader will be able to apply these improvements in the power inverters to his or her problems for high-performance power inverters.

To facilitate this goal, Chapter 1 introduces "minimum switch" converter topologies and control of passive load as well as split-single-phase induction motor so that the total harmonic distribution (THD) of the inverter is improved. Two-stage connection including simple VSI inverter and center-tapped series-parallel resonant LC filter (LCL2C2) both with neutral point and half-bridge matrix converter has been applied in the two-phase inverters. Later one includes one-leg half-bridge matrix converter and the AC neutral point network as a new type of converter with two-phase outputs loading the resistive-inductive or motoric loads. Besides, the running capacitor creating needed phase shift (90 deg.) has been electronically switched due to varied load. Moreover, analysis and modeling of such a new type of single-leg AC/AC converter with two-phase outputs have been done. The proposed inverter and converter topologies have been simulated by MATLAB/Simulink and verified in LT-Spice environment. Combination of tuned LC filter and switched capacitor brings a good quality of output quantities of converter and represents the main contribution of the chapter.

This is followed by Chapter 2, where authors propose a new design of multilevel inverter configuration to reduce the component count and improve the quality of waveform in a photovoltaic system. The proposed configuration operates at the binary asymmetric condition to generate high output voltage level with small amount of harmonic distortion. Unipolar trapezoidal reference with triangular carriers has been used in the proposed inverter to produce the desired switching pulses and generate the required output voltage level. Moreover, separate DC sources of proposed configuration are replaced by the array of photovoltaic panels to check the configuration with the renewable energy source. Finally, to show the effectiveness of the proposed configuration, an experimental setup is implemented.

In Chapter 3, authors review the cascaded H-bridge based on current source inverter topology. The first description of the power topology has been presented from the point of view of the current source single-phase inverter and its connection in series with other inverters. Then, modulation of the single-phase inverter has been studied, and the use of multilevel modulation techniques and their use in the proposed power topology have been reviewed and simulated. Next, key design guidelines of the output capacitor and the DC inductor have been reviewed. Finally, an application example for AC drives simulated in PSIM has been presented. From the study, it can be concluded that the main advantages of the topology are the quality of both input currents and load voltage, while its main drawback is the use of a bulky DC inductor because of the use of current source inverters and the oscillating power drained by the inverter from the DC side. In the same way as classic cascaded H-bridge topologies, the use of the proposal topology allows to use semiconductors and passive component with lower voltage and current rating that are required by the load.

Authors of Chapter 4 present PDM control on the series resonance inverter, which led to ac-ac converters with high efficiency (zero switching loss), small size (no storage capacity), and the possibility of a self-power factor correction. The PDM control joins together between the concepts of soft switching and hard switching. Due to the complexity of the operating analysis for these converters, the average modeling facilitates the analysis of the operation and leads to establish (i) an analytical expression of the power factor, (ii) the linearity conditions of the power characteristic, and (iii) a model of ac-ac series resonant multiconverter, which is independent of the carriers. In the case of ac-ac series resonant multiconverter, the coordination of carriers allows to shape the power characteristic. Among the three types of coordination presented, there is an original coordinate that linearizes the power characteristic. The results have been validated by simulations carried out in Matlab SimPower systems.

A general study of the dual-inverter topology for supplying open-ended loads is discussed in Chapter 5, where the authors study a type of connection consisting on leaving both terminal ends of the load open as an alternative to standard wye or delta connection. To supply loads with this type of connection, two power inverters (one at each terminal end of the load) are required in a circuit topology called dual inverter. In this chapter, the advantages and issues of such converter have been studied, and different modulation strategies have been shown and discussed. Moreover, multilevel dual-inverter converters have been presented as an extension to the basic two-level idea. For evaluation purposes, simulation results have been presented.

The objective of Chapter 6 is to put into evidence the teaching aspects through the applicative research in the field of the electric drives. In this chapter, the author provides the basic and advanced aspects of the electric drives control based on the most used electrical machine: three-phase induction motor (IM). The research work has been presented in didactical way, starting with the conventional vector control, followed by the integration of the model reference adaptive control into the specific IM-based drive. The verified numerical simulation results push the research process through the implementation way. In order to increase the IM drives' efficiency, the real-time implementation of the most used modulation techniques has been provided. Based on the dSpace platform, interfaced by Control Desk, the experimental results have been obtained. Both the performances of the cascaded control and model reference adaptive control have been shown.

Ali Saghafinia
Department of Electrical and Computer Engineering
Isfahan University of Technology
Isfahan, Iran

Chapter 1

Two-Phase Inverters with Minimum Switching Devices

Branislav Dobrucky, Tomas Laskody and
Roman Konarik

Additional information is available at the end of the chapter

http://dx.doi.org/10.5772/67743

Abstract

The chapter deals with two-phase inverters with minimum switching devices whereby the main emphasis is devoted to 'minimum switches' converter topologies and control of passive load as well as split-single-phase induction motor. Such a converter consists of one-leg half-bridge matrix converter and the ac neutral point network as a new type of converter with two phase outputs loading the resistive-inductive or motoric loads. As harmonic content of the voltage of both phases gives very high value of total harmonic distortion (THD), roughly 86%, the current waveforms should be improved by using serial LC filter that brings much more suitable value of THD. Besides, the running capacitor creating needed phase shift (90°) is electronically switched due to varied load. Analysis and modeling of such a new type of single-leg ac/ac converter with two phase outputs are done. The proposed topologies were simulated by Matlab/Simulink and verified in an LT spice environment. Worked-out simulation results are in good agreement with theoretical assumptions and make possible to give recommendation for the fair and right design of the chosen type of converter. Combination of mentioned measures brings a good quality of output quantities of converter and represents the main contribution of the chapter.

Keywords: two-phase inverter, matrix converter, one-leg VSI converter, LC resonant filter, half bridge connection, bidirectional switch, modeling, LT Spice simulation

1. Introduction

In spite that the generation and transmission of electric power are done by means of the three-phase ac system, today, the development of the two-phase system is still continued mainly to split-single-phase IM motor supply that is documented in this chapter and given references. So, the first venture into the realm of polyphase electric power has used only two alternating current phases rather than three but with pulsating power flow to motor in contrast to constant power of three-phase system [1]. In regard to topologies of the two-phase inverters, mostly three-leg ones with six switches or two legs with four switches are used. Evaluation of low-cost topologies for the two-phase induction motor (IM), which drives in an industrial application, is analyzed and discussed in Refs. [2–4]. Half-bridge two-phase voltage source inverters (VSI) for two-phase (IM) supply are described in Refs. [5–8]. Besides, there exists also a possibility to supply three-phase induction motor by the two-phase inverter [9].

Regarding to minimum switching devices, two-phase one-leg VSI inverters for the two-phase IM supply, there are works of Chomat et al. in Refs. [10–12]. In those, the operation of the motor at nominal frequency is different from the reduced frequency operation when phase shift of auxiliary phase is provided by a capacitor. Due to variable load, it is useful to change the value of capacitance in auxiliary phase, so the electronically switched capacitor techniques are used [13, 14]. Current pulse-width-modulation (PWM) or space vector modulation (SWM) provides demanded sinusoidal waveform and feedback control [15, 16].

A new original topology of single-leg direct matrix converter was first published by authors of this chapter in Ref. [17] based on the works in Refs. [10, 18]. The number of switches is minimized, but total harmonic distortion (THD) of auxiliary-phase voltage is very high (86% at 50 Hz, 69% at 33.33 Hz) and consequently current distortions too (68 and 43%, respectively). Therefore, the improved two-phase one-leg matrix converter is completed by an LC filter [19] designed by Dobrucký et al. [20]. Combination of tuned LC filter and switched capacitor brings a new quality of output quantities of converter, which provides acceptable THD and makes possible field-oriented control (FOC) of the IM motor.

The chapter is organized as follows. First, the basic topologies of one-, two-, three-, and four-leg VSI inverters for two-phase application are described. Next, the special topologies using matrix inverters for two-phase application are introduced. Possibilities of use of switched capacitor for auxiliary circuit phase control providing the use of LC filter are described, and simulation study of Matlab/Simulink and LT Spice with passive RL and active motoric loads are given. Afterward, current controlled PWM (or hysteresis control) is worked out, and finally, conclusion is described.

2. Basic topologies of VSI inverters for two-phase application

2.1. Two-phase voltage source inverter with two legs

The topology (**Figure 1**) consists of four semiconductor switches. A low number of a semiconductor switches is the main advantage of the topology. Those switches create two half-bridge inverters, each of them powers one of the windings. The disadvantage, which this topology suffers from, is low magnitude of the output voltage. It is half of an interlink DC voltage in Refs. [2, 21]. Another disadvantage is hidden in control of the switches, which is only bipolar PWM can be used [6], which has further negative consequences.

It is also possible to supply three-phase IM motor by two-phase VSI inverter [9], **Figure 2**.

http://dx.doi.org/10.5772/67743

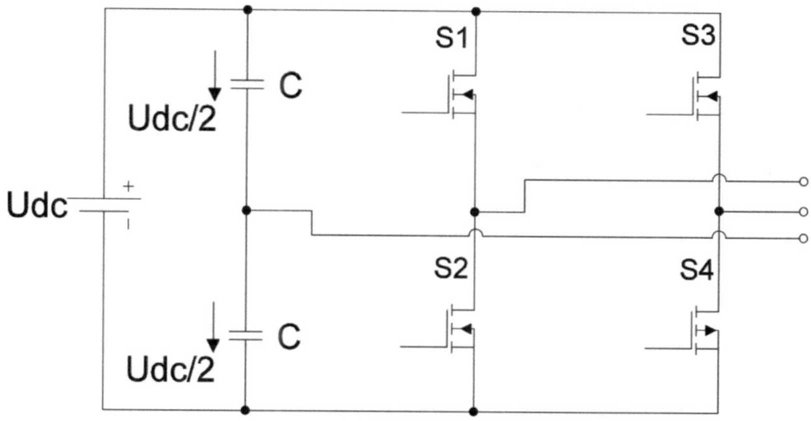

Figure 1. Voltage source inverter with two legs.

In opposite, there is also a possibility to supply two-phase IM by three-phase three-leg inverter (at the next).

2.2. Two-phase voltage source inverter with three legs

The topology shown in **Figure 3** consists of six semiconductor switches. Two of the three-leg are used for the power supply of the motor windings and third leg is used for creation of common phase of the motor [2, 21]. As a control of the switches, the modified SPWM [6, 22] can be used to describe the use of sin(ωt), cos(ωt), and –sin(ωt) as the reference voltages. The advantage compared to the converter with two legs is in better usage of a DC interlink. While inverter with two legs can put only half of DC voltage interlink magnitude, inverter with three legs is able to use $U_{DC}/\sqrt{2}$.

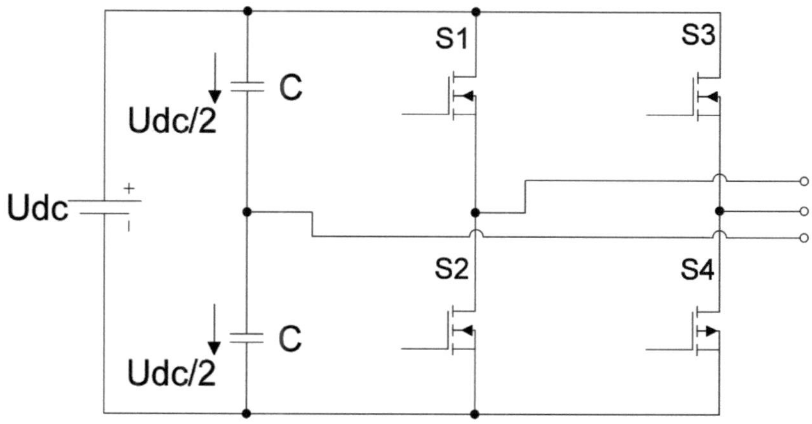

Figure 2. Voltage source inverter with two legs for supply of three-phase IM.

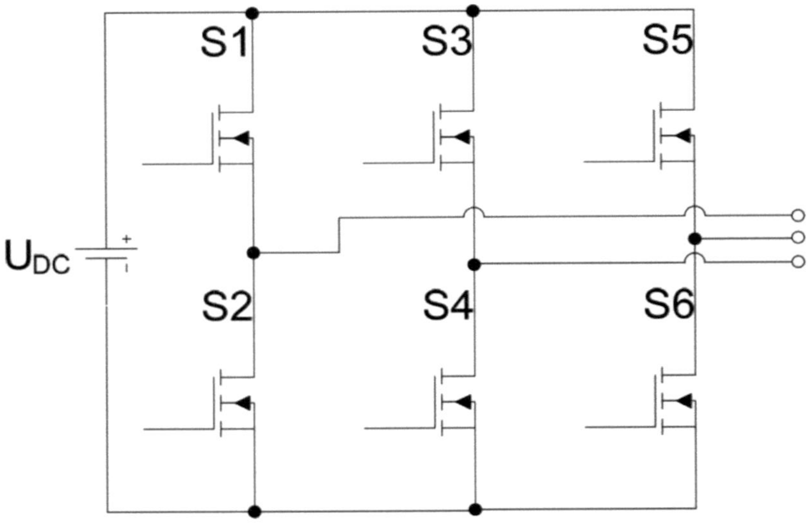

Figure 3. Voltage source inverter with three legs.

2.3. Two-phase voltage source inverter with four legs

Another possible topology that can be used to fed the two-phase induction motor (**Figure 4**) is created by eight switches. Each phase is fed by one full-bridge inverter.

The topology uses a larger amount of switches (eight ones), and therefore, the topology is able to use entire magnitude of DC interlink voltage [7, 21].

Model of two-phase IM motor is well known [7–9, 13, 14]. So, the electric machine being considered may be described by the following set of ordinary differential equations in the $\alpha\beta$-stator reference coordinate frame under the commonly used simplifying assumptions:

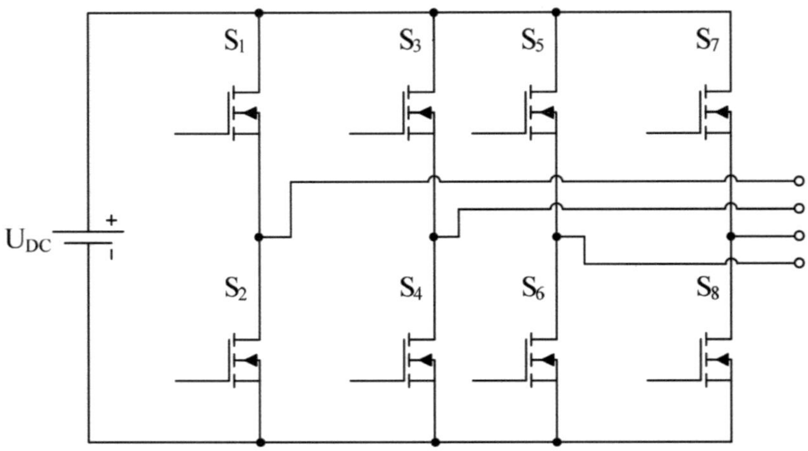

Figure 4. Voltage source inverter with four-leg and four-output terminals.

http://dx.doi.org/10.5772/67743

$$u_{s\alpha} = R_{s\alpha} i_{s\alpha} + L_{s\alpha} \frac{di_{s\alpha}}{dt} + L_{M\alpha} \frac{di_{r\alpha}}{dt}, \tag{1}$$

$$u_{s\beta} = R_{s\beta} i_{s\beta} + L_{s\beta} \frac{di_{s\beta}}{dt} + L_{M\beta} \frac{di_{r\beta}}{dt}, \tag{2}$$

$$0 = R_{r\alpha} i_{r\alpha} + L_{r\alpha} \frac{di_{r\alpha}}{dt} + L_{M\alpha} \frac{di_{s\alpha}}{dt} + + \frac{1}{N} \omega_m (L_{r\beta} i_{r\beta} + L_{M\beta} i_{s\beta}), \tag{3}$$

$$0 = R_{r\beta} i_{r\beta} + L_{r\beta} \frac{di_{r\beta}}{dt} + L_{M\beta} \frac{di_{s\beta}}{dt} + - N\omega_m (L_{r\alpha} i_{r\alpha} + L_{M\alpha} i_{s\alpha}), \tag{4}$$

$$T_e = pp \left[N(L_{r\alpha} i_{r\alpha} + L_{M\alpha} i_{s\alpha}) i_{r\beta} - \frac{1}{N} (L_{r\beta} i_{r\beta} + L_{M\beta} i_{s\beta}) i_{r\alpha} \right], \tag{5}$$

$$T_e = T_{\text{load}} + J \frac{d\omega_m}{dt}. \tag{6}$$

where N is the ratio between the effective numbers of turns in the auxiliary and the main stator windings, ω_m is the mechanical angular speed, and pp is the number of pole pairs.

As control methods, it can be used modern control ones: field-oriented vector control as well as space vector pulse width modulation [6, 15]. Some results of operation of two-phase IM supply are shown in **Figure 5a**, start-up and steady state (b).

The topology VSI with two legs controlled by SVPWM is able to turn on a four active voltage vector but not able to turn on a zero voltage vector, which is its major disadvantage [21].

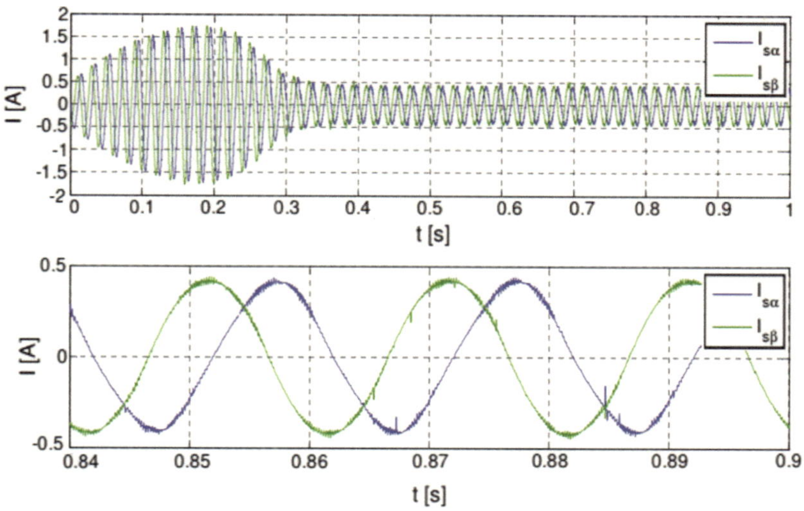

Figure 5. Behaviour of two-phase IM supplied by two-phase VSI inverter; (top) start-up and (bottom) steady-state operation [21].

Substituting VSI topology by matrix one will be able to turn on eight active voltage space vectors with turn-on times as shown in **Figure 6 (top)** but still no zero space vectors, **Figure 6 (bottom)**.

Instead of VSI inverters with two, three, and four legs, there can be used matrix converter [21], but the number of switching devices is rather higher, nearly two times.

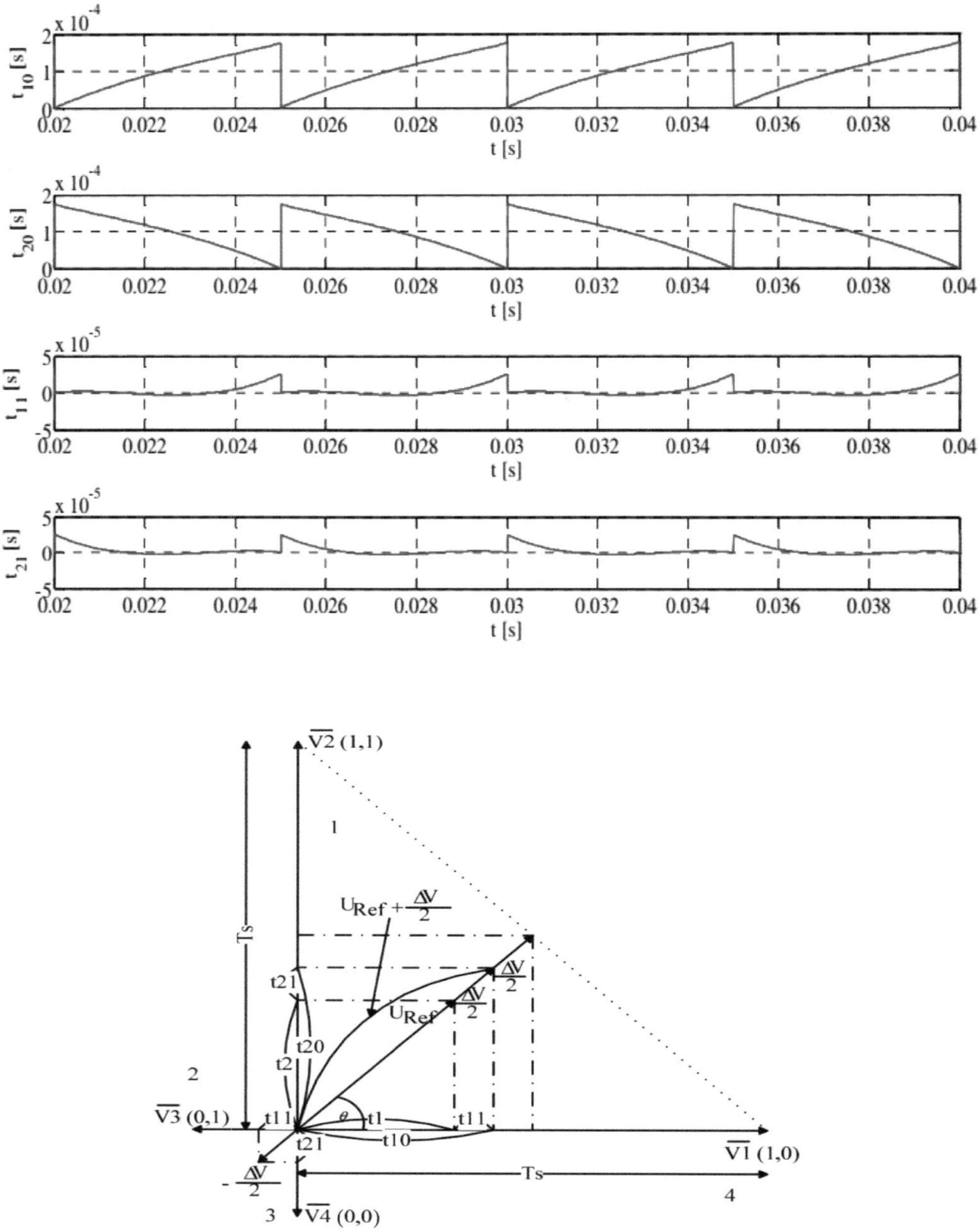

Figure 6. Two-leg VSI; (top) turn-on times of active voltage vectors and (bottom) creating SVPWM [21].

3. Two-phase inverters with minimum switching devices

3.1. Two-phase voltage source inverter with one leg

Minimum of switching devices: two switches for inverter, two diodes for rectifiers, are reached by the one-leg VSI inverter [8, 10], **Figure 7**.

Anyway, it also needs two antiparallel diodes and two bulky capacitors. Schematic of VSI in **Figure 7** is dedicated for ac motors. In full speed operation, the one leg of VSI with switches provides phase shift of 90°, since in reduced speed operation, the shift is created by a capacitor.

3.2. Two-phase voltage matrix converter with one and two legs

Instead of one-leg VSI inverters that can be used matrix converters is based on single-phase matrix converter. The matrix converter has some specific advantages over voltage source inverter in the size of the device, the lack of intermediate circuit, and also reduction in needed capacity [18]. The disadvantages are higher cost and also higher number of switching elements. Matrix converter that consists of just one single leg is described as original one, in Section 4.

3.3. Two-phase LCL2C2 inverter with two-leg matrix converter

One possibility for the first stage is to use a resonant converter, for instance an LCL2C2 resonant converter, **Figure 8a**. The second stage can be created as a two-leg two-phase matrix converter. This resonant converter consists of four bidirectional switches. For his control, an SVPWM modulation can be used, which is very like of SVPWM modulation for two-phase two-leg voltage VSI inverter.

The difference is the need to monitor the voltage polarity in an intermediate circuit and properly toggle the combination of active vectors. Unlike the two-phase two-leg voltage source

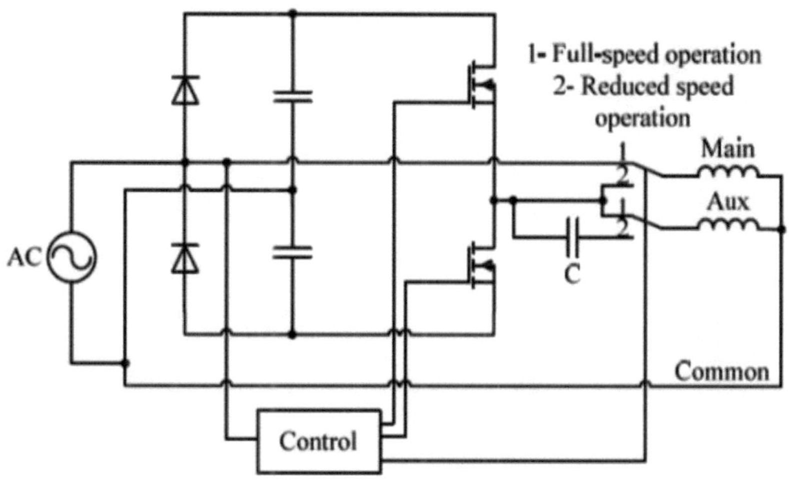

Figure 7. Schematic of one-leg VSI inverter [10].

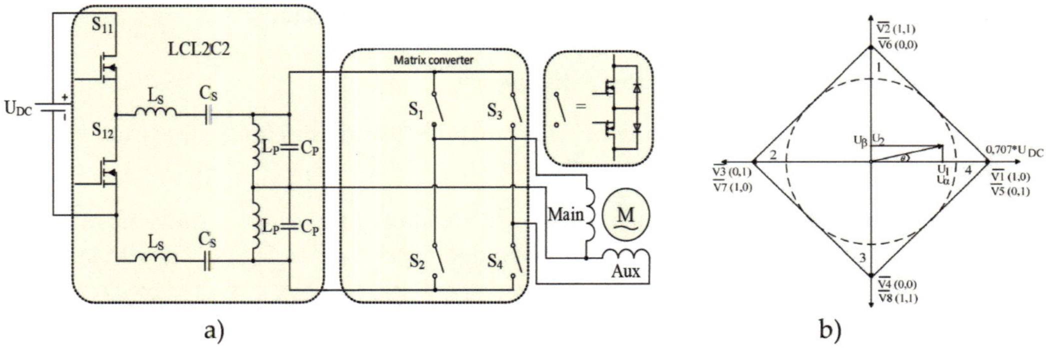

Figure 8. Two-stage MxC; (a) with resonant converter as a first stage and (b) its space vectors [21].

inverter, the two-phase two-leg matrix converter has double number of active vectors, **Figure 8b**. It is necessary to switch on the active vectors V1–V4 when the voltage in intermediate circuit is positive. If there is a negative voltage in the intermediate circuit, vectors V5–V8 are switched on. Reference voltage for the switches is shown in **Figure 9**.

The operation of matrix converter with motoric load in an open-loop operation and detail of stator currents and adequate stator voltages during two periods at steady state are shown in **Figure 10**.

Anyway, number of switching devices using two-phase LCL2C2 inverter with two-leg matrix converter is still high (2 + 4 that means six switches).

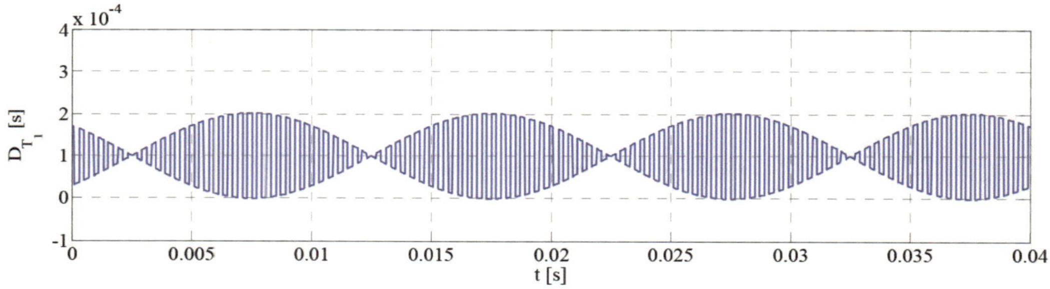

Figure 9. Waveforms of reference voltages for two-phase matrix converter [21].

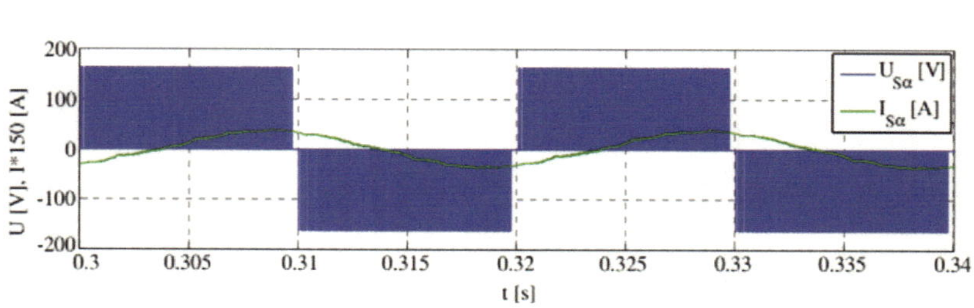

Figure 10. Stator voltages and stator currents of α-phase during two periods [21].

4. Single-leg topology using matrix converters for two-phase application

Another possibility how to reduce number of switching devices is presented by the special connection of one-leg matrix converter supplied direct from the network. As new type of two-phase converters using matrix converter for two-phase applications has been developed for single-leg matrix converter [17].

4.1. Basic topology of single-leg matrix converter

A novel supply system for two-phase induction motor by a single-leg matrix converter was introduced in work [17] using principle of *single-phase matrix converter where one phase is substituted by harmonic network voltage*, **Figure 11a** and **b**.

In the circuit operation of full speed regime, the voltage of auxiliary phase is possible to express as

$$u_{\text{aux}}(t) = U_M \text{sign}[\sin(\omega t)]\text{abs}[\cos(\omega t)]. \tag{7}$$

For reduced speed, the voltage of both main and auxiliary phases is expressed as

$$u_{\text{mxc}}(t) = U_M \text{sign}[N \sin(\omega t)]\text{abs}[\sin(\omega t)], \tag{8}$$

$$N = \frac{f_{mxc}}{f_{ac}} = \frac{T_{ac}}{T_{mxc}}. \tag{9}$$

The input and output waveforms of MxC in full speed mode and reduced speed mode are shown in **Figures 12** and **13**. The switching logic of the control system that creates the desired output voltage from input voltage is shown in **Figure 14**.

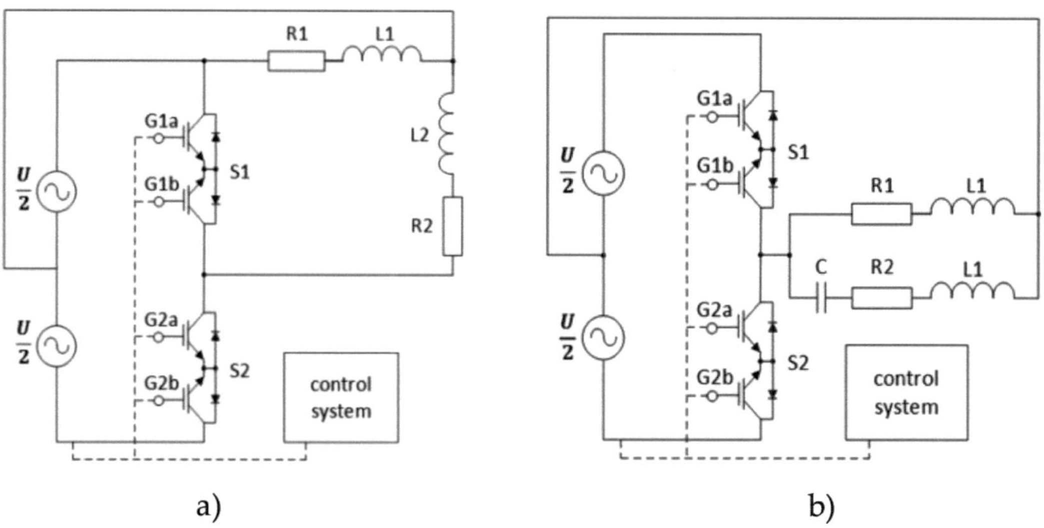

Figure 11. One-Leg MxC; (a) schematics for full speed at 50 Hz and (b) for reduce speed (<50 Hz) [17].

Among various questions, the first question is how value of the fundamental harmonic of auxiliary phase will reached. Using Fourier analysis of the one fourth of the waveform, **Figure 15**, one can write equations:

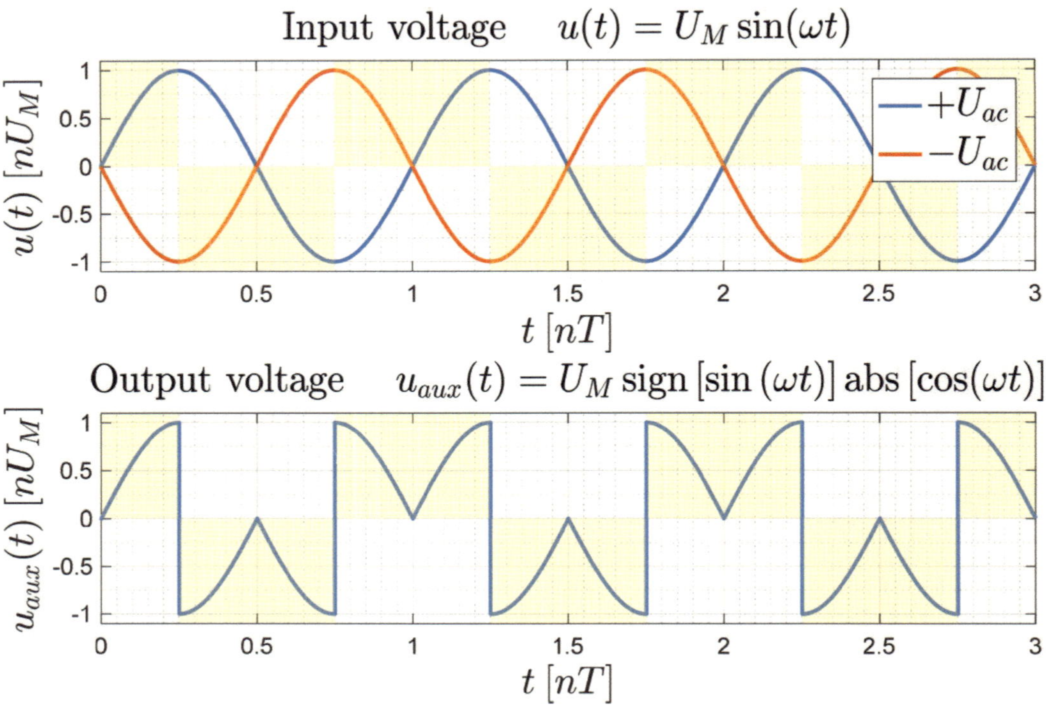

Figure 12. Voltages of SLMxC in full speed mode 50 Hz; (top) input network voltage and (bottom) SLMxC output voltage.

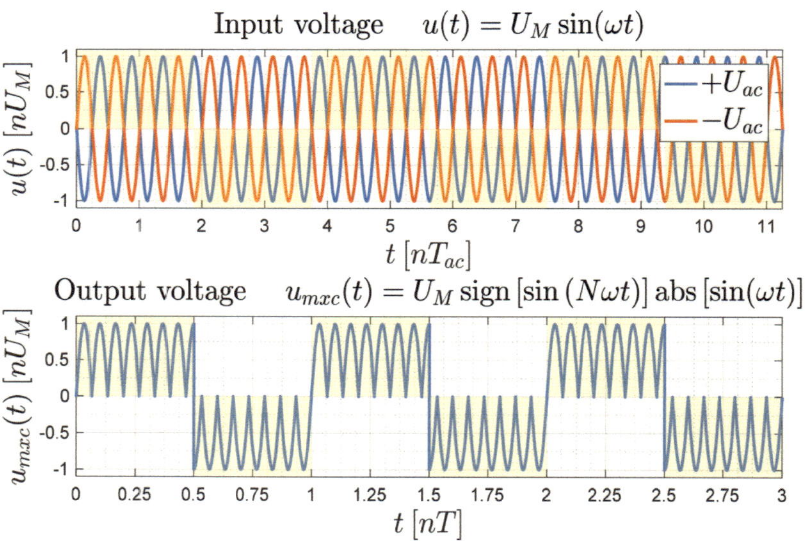

Figure 13. Voltages of SLMxC in reduced speed mode 6.66 Hz; (top) input and (bottom) output voltages.

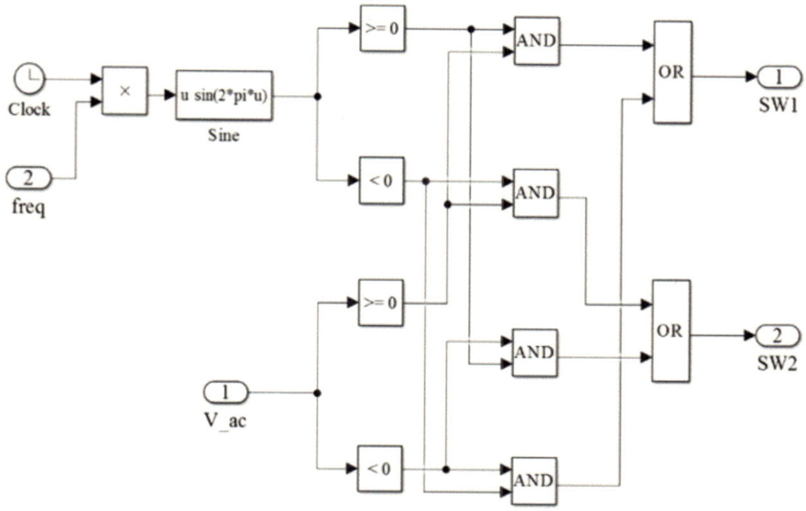

Figure 14. The switching logic for main and auxiliary phases.

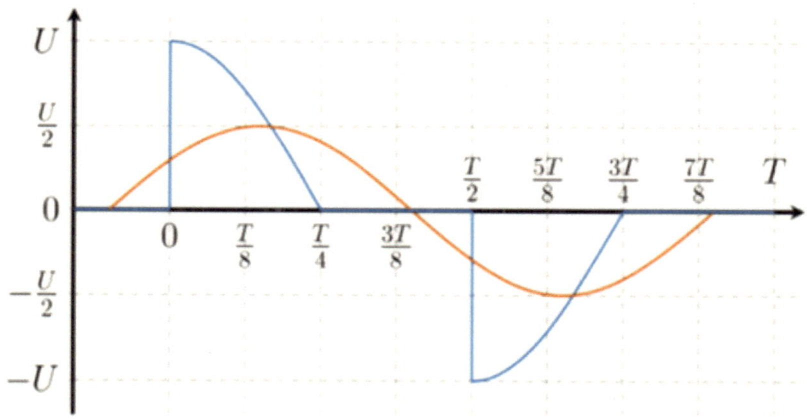

Figure 15. To harmonic analysis for the first part of the waveform.

Definite relations

$$f_1(t) = a_1 \cos(\omega_1 t) + b_1 \sin(\omega_1 t) = A_1 \sin(\omega_1 t + \varphi_1), \tag{10}$$

$$A_1 = \sqrt[2]{a_1{}^2 + b_1{}^2};\ \varphi_1 = \text{arctag}\frac{b_1}{a_1}, \tag{11}$$

$$a_1 = \frac{2}{T}\int_0^T f(t)\cos(\omega_1 t)\mathrm{d}t;\ b_1 = \frac{2}{T}\int_0^T f(t)\sin(\omega_1 t)\mathrm{d}t. \tag{12}$$

Then

$$a_1 = \frac{2}{T}\int_0^{\frac{T}{4}} \cos^2(\omega_1 t)\mathrm{d}t = \frac{2}{T}\int_0^{\frac{T}{4}} \frac{1}{2}[1 + \cos(2\omega_1 t)]\mathrm{d}t$$
$$= \frac{1}{T}\int_0^{\frac{T}{4}} [1 + \cos(2\omega_1 t)]\mathrm{d}t = \frac{1}{T}[t]_0^{T/4} + \frac{1}{T}\frac{1}{2\omega_1}[\sin(2\omega_1 t)]_0^{T/4} = \frac{1}{4}, \tag{13}$$

$$b_1 = \frac{4}{T}\int_0^{T/2} f(t)\sin(\omega_1 t) = \frac{2}{T}\int_0^{\frac{T}{4}} \cos(\omega_1 t)\sin(\omega_1 t)\mathrm{d}t$$
$$= \frac{2}{T}\int_0^{\frac{T}{4}} \frac{1}{2}[\sin(2\omega_1 t)]\mathrm{d}t = \frac{1}{T}\frac{1}{2\omega_1}[-\cos(2\omega_1 t)]_0^{T/4} = \frac{1}{4\pi}[\cos(2\omega_1 t)]_{T/4}^{0} = \frac{1}{2\pi}. \tag{14}$$

Fundamental harmonic waveform

$$u_1(t) = \frac{1}{4}\cos(\omega_1 t) + \frac{1}{2\pi}\sin(\omega_1 t) = \sqrt[2]{\left(\frac{1}{4}\right)^2 + \left(\frac{1}{2\pi}\right)^2}\sin\left(\omega_1 t + \mathrm{arctag}\frac{1/2\pi}{1/4}\right). \tag{15}$$

After calculation

$$u_1(t) = 0.296\sin(\omega_1 t + 32.48°). \tag{16}$$

The value if fundamental harmonic at the middle of half-period

$$A_1|_{\frac{\pi}{2}} = 0.296\cos(32.48°) = 0.249. \tag{17}$$

The contribution from the second part of auxiliary-phase waveform will be the same, **Figure 16**. So, this means that maximal magnitude of auxiliary-phase fundamental harmonic is

$$2A_1|_{\frac{\pi}{2}} = 2 \times 0.249 = 0.498 \approx 0.5. \tag{18}$$

Thus, the RMS value of the output voltage of the one-leg converter should be two times greater than requested voltage of the main phase of the system.

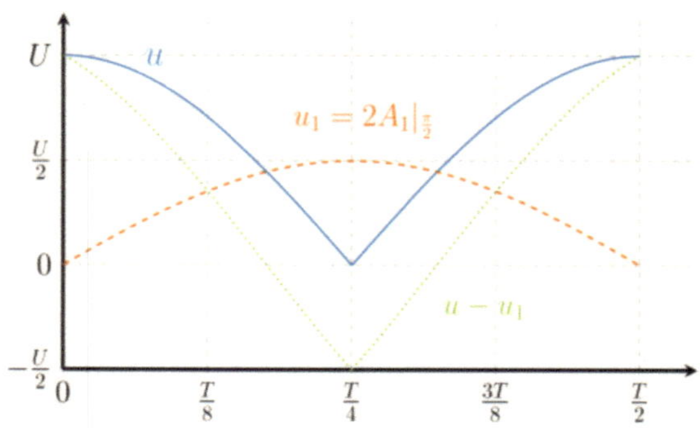

Figure 16. Auxiliary-phase waveforms; (solid) total wave, (dashed) fundamental harmonic and (dotted) sum of higher harmonics.

Basic scheme of MxC converter for reduced speed is different from full speed and given in **Figure 11b**. The phase shift of auxiliary phase is provided by the capacitor C_{aux}. Vector diagram for auxiliary-phase impedances is given in **Figure 17**.

By calculating C_{aux} for 'quadratic mean' frequency band 33.33 Hz from vector diagram, the capacitor value for auxiliary phase can be determined

$$|Z_{aux}| = |Z_{main}|,\ f_{qm} = \sqrt{\frac{f_{lo}^2 + f_{hi}^2}{2}}, \tag{19}$$

$$\left|\frac{1}{\omega C_{aux}}\right| = |Z_{aux}|\cos\varphi + |\omega L_2|, \tag{20}$$

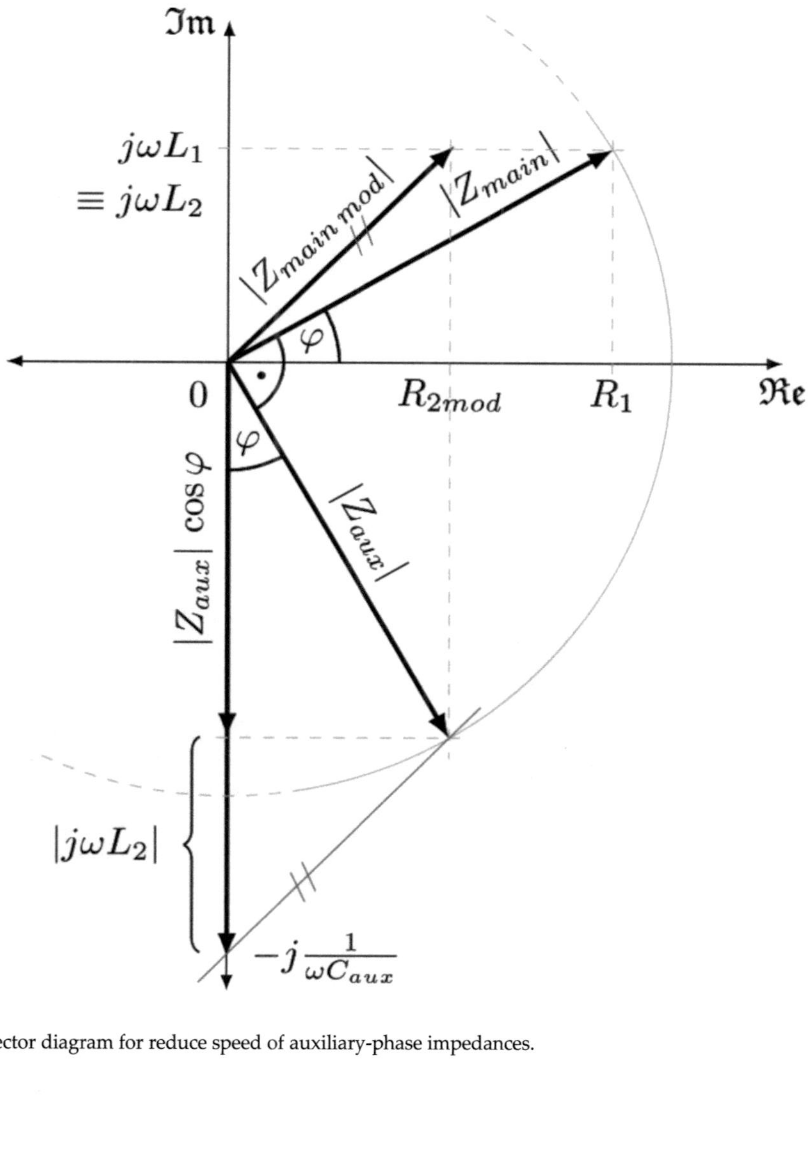

Figure 17. Vector diagram for reduce speed of auxiliary-phase impedances.

$$C_{aux} = \frac{1}{\omega} \frac{1}{|Z_{aux}|\cos\varphi + |\omega L_2|}. \tag{21}$$

There is equality of $|\omega L_2| = |\omega L_1|$ to be the same magnetic flux in both main and auxiliary phases.

Analytical differential equations for main- and auxiliary-phase state-space variables (**Figure 11b**) are:

$$\frac{di_{main}}{dt} = -\frac{1}{\tau} i_{main} + \frac{1}{\tau} \frac{U_m}{R_1} \sin(\omega t), \tag{22}$$

$$\frac{di_{aux}}{dt} = -\frac{1}{\tau} i_{aux} + \frac{1}{L_1} u_C + \frac{1}{\tau} \frac{U_m}{R_2} \sin(\omega t), \tag{23}$$

$$\frac{du_C}{dt} = \frac{1}{C} i_{aux}, \tag{24}$$

where τ is time constant accordingly R_1 and R_2 resistances.

After time discretization (e.g. using Euler's method), we obtain discrete dynamic model suitable for simulation in Matlab/Simulink.

Similar to one-leg VSI inverter [8, 12], the number of switches of single-leg MxC is minimized but total harmonic distortion auxiliary-phase voltage is very high (86% at 50Hz, 69% at 33.33 Hz) and consequently current distortions too (68% and 43%, respectively), see **Figure 18** in Section 6 for simulation results where **Figure 18a** is for 50 Hz and **Figure 18b** is for 33.33 Hz.

4.2. Single-leg matrix converter supplying IM without LC filter and PWM

Due to very high THD of the main and auxiliary voltages and current, there is a problem regarding to electromagnetic torque generated by two-phase IM motor. Therefore, neither the operation at 50 Hz nor at reduced frequency (33.33 or 25 Hz), under nominal torque and during start-up, could be provided successfully.

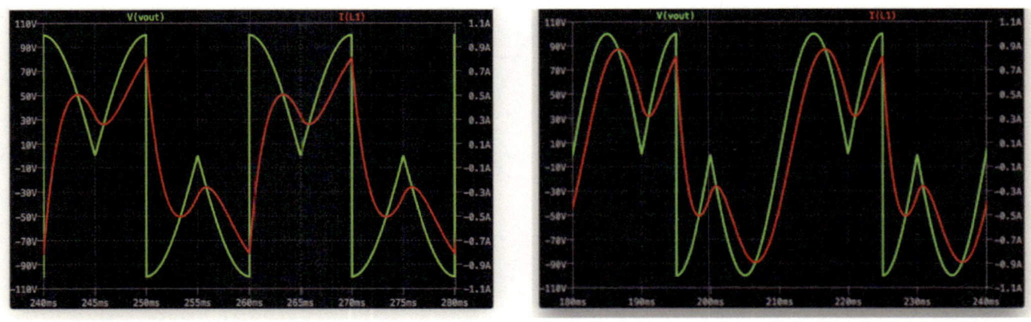

a) b)

Figure 18. One-Leg MxC; (a) voltage and current of auxiliary-phase at 50 Hz and (b) at 33.33 Hz.

http://dx.doi.org/10.5772/67743

Thus, we have to accept some measures for successfully operating PWM control, adding LC resonant filter, and/or using switched capacitor.

Parameters of the two-phase IM used for simulation are therefore given in Section 5 with PWM controlled motor.

4.3. Improving single-leg matrix converter using LC filter

Thus, it is necessary to improve current waveforms. Current-controlled PWM modulation (CC-PWM) for full-speed operation is not possible to use because of decreasing of auxiliary-phase voltage. Other possibility that has been used is using of LC resonant circuit that can be used both in auxiliary and main-phase circuits, **Figure 19a** and **b**.

Analytical differential equations for main- and auxiliary-phase state-space variables (**Figure 19b**)

$$\frac{di_{\mathrm{main}}}{dt} = -\frac{1}{\tau} i_{\mathrm{main}} + \frac{1}{L_1} u_{\mathrm{Cres}} + \frac{1}{\tau}\frac{U_m}{R_1}\sin(\omega t), \tag{25}$$

$$\frac{di_{\mathrm{aux}}}{dt} = -\frac{1}{\tau} i_{\mathrm{aux}} + \frac{1}{L_{2\mathrm{mod}}} u_C + \frac{1}{\tau}\frac{U_m}{R_{2\mathrm{mod}}}\sin(\omega t), \tag{26}$$

$$\frac{du_{\mathrm{Cres}}}{dt} = \frac{1}{C_{\mathrm{res}}} i_{\mathrm{main}}, \tag{27}$$

$$\frac{du_C}{dt} = \frac{1}{C_{\mathrm{aux}}} i_{\mathrm{aux}}, \tag{28}$$

where τ is time constant accordingly to R_1 and R_2 resistances, and C_{aux} is series connected C and C_{res} capacitors.

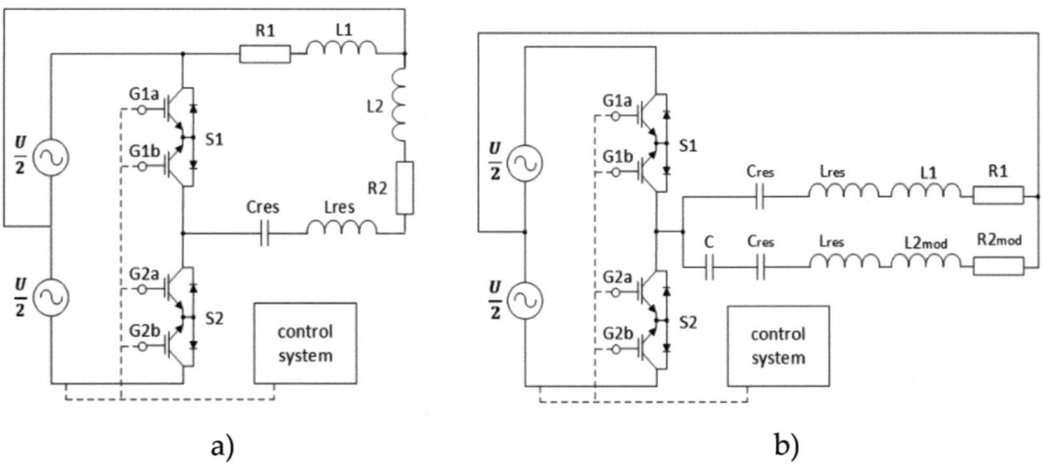

Figure 19. Schematics of SLMxC; (a) for full speed at 50 Hz and (b) for reduce speed (<50 Hz) with LC circuits.

After time discretization (e.g. using Euler's method), we obtain discrete dynamic model suitable for simulation in Matlab/Simulink.

After realization of above measures, the total harmonic distortion of main and auxiliary-phase currents will be much better such as 12.47% at 50 Hz and 8.47% at 33.33 Hz with LC circuits, see **Figure 20a** for 50 Hz and **Figure 20b** for 33.33 Hz in Section 6. By suitable design of LC elements [20], it is possible to reach the best value of main-and auxiliary-phase current THDs (<5%) but the size of the LC elements will be rather high.

Resonant LC filter can be tuned either for given frequency 50 Hz, schematic in **Figure 19a**, or for 'quadratic mean' frequency band 33.33 Hz, schematic in **Figure 19b**. Design of LC filter has been done using design procedure by Dobrucký et al. [20].

4.4. Improving single-leg matrix converter by combining switched capacitors and LC filter

Auxiliary-phase advancing using switched C or L elements can be provided by different ways [13, 14]

- switched capacitor with four switches networks, **Figure 21a**
- switched capacitors with two switches and two capacitors, **Figure 21b**
- switched RLC (inductor and capacitor) circuit with two switches connected as series network, **Figure 21c**
- switched inductor with two switches connected as parallel network, **Figure 21d**

Since first two possibility items operate with full controlled bidirectional switches the later two can be operated by ordinary thyristors with uncontrolled switching-off of the circuit current.

A periodically reversed switched capacitor is connected in series with an RL load supplied from a sinusoidal voltage source. To control the phase of the fundamental component of load current, a suitable algorithm for the switching of the capacitor is derived and tested. The operation of the switch pairs is complementary and supports a pulse width modulated regime where the duty factor of the dominant pair, e.g. S1, is restricted to vary between 0.5 and unity.

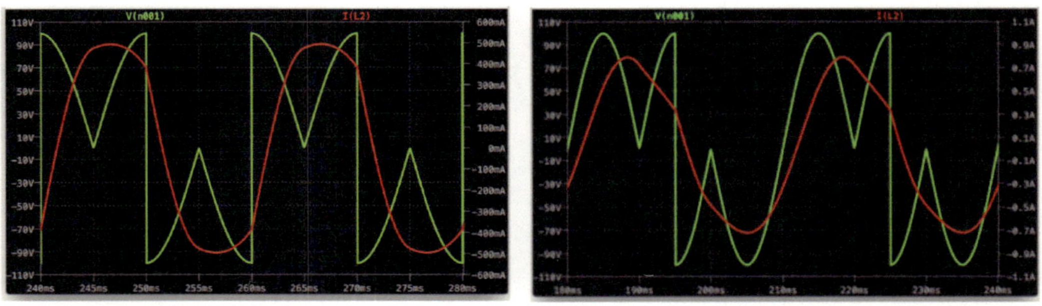

a) b)

Figure 20. Voltage and current of auxiliary-phase at 50 Hz (a) and at 33.33 Hz (b) with additional LC circuit.

http://dx.doi.org/10.5772/67743

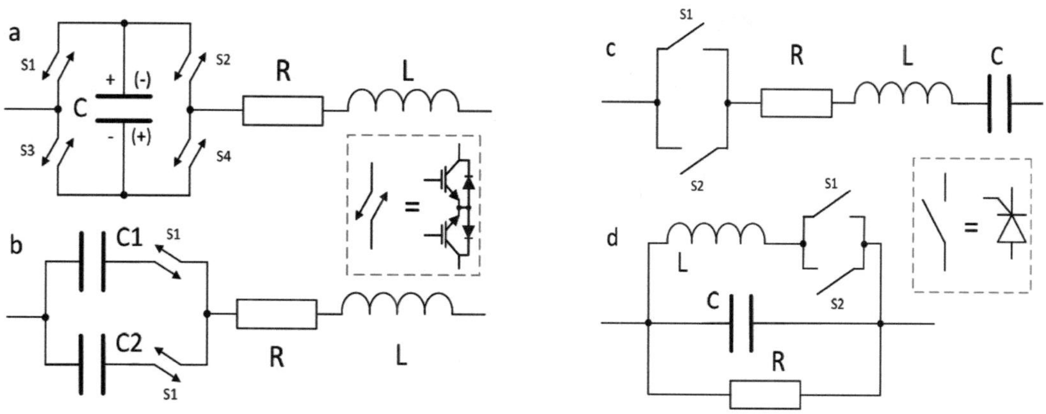

Figure 21. Possibilities of switched capacitor networks for phase shift control.

We have chosen a variant shown in **Figure 21b** because of variant **Figure 19a** that needs high number of switches and variants **Figure 19c** and **d** operate with uncontrolled switching-off of the circuit current only, with an auxiliary winding through the bidirectional choppers S1 and S2 controlled by PWM with a frequency of about 1 kHz. The capacitance is changed by variation of the duty cycle

$$D_{\mathrm{tc}} = \frac{t_{\mathrm{on}}}{t_{\mathrm{on}} + t_{\mathrm{off}}}, \tag{29}$$

from 0 to 1.

The switches S1, S2 are linked with each other by an inverted logic. One capacitor has high capacitance and the other has low capacitance. The desired capacitance is set as a function of duty cycle for the switching between these two capacitors.

The equation for the controlled switched capacitance C can be derived from the energy stored in the capacitors [8]

$$E = \frac{1}{2} C U^2. \tag{30}$$

From Fourier analysis, the dc component of periodic waveform is equal to its average value

$$U_{\mathrm{s}} = \frac{1}{\mathrm{T}} \int_0^T \mathrm{u(t)d}t = D_{\mathrm{tc}}\, U, \tag{31}$$

where $\mathrm{D_{tc}}$ is the duty cycle and T is the switching period. Therefore, the average voltages on the capacitors are

$$U_1 = D_{\mathrm{tc}}\, U,\ U_2 = (1 - D_{\mathrm{tc}})\, U. \tag{32}$$

Then, the total energy stored in the two capacitors is

$$\frac{1}{2}CU^2 = \frac{1}{2}C_1U_1^2 + \frac{1}{2}C_2U_2^2. \tag{33}$$

Thus,

$$CU^2 = C_1(D_{tc}U)^2 + C_2[(1 - D_{tc})U]^2. \tag{34}$$

After further simplification, we get the final equation for the switched capacitance as the function of duty cycle

$$C = C_1{D_{tc}}^2 + C_2(1 - D_{tc})^2. \tag{35}$$

The absolute value of the auxiliary impedance $|Z_{aux}|$ and the phase angle φ used in the function calculations are given as

$$|Z_{aux}| = \sqrt{R_{2mod}^2 + (\omega L_2)^2}, \tag{36}$$

$$\varphi = \tan^{-1}\frac{\omega L_2}{R_2}. \tag{37}$$

Combining Eq. (21) for the auxiliary-phase capacitance C_{aux} with Eq. (35) for switched capacitance $C_{aux} = C$ yields

$$0 = (C_1 + C_2){D_{tc}}^2 - 2C_2D_{tc} + C_2 - C_{aux}(\omega), \tag{38}$$

and

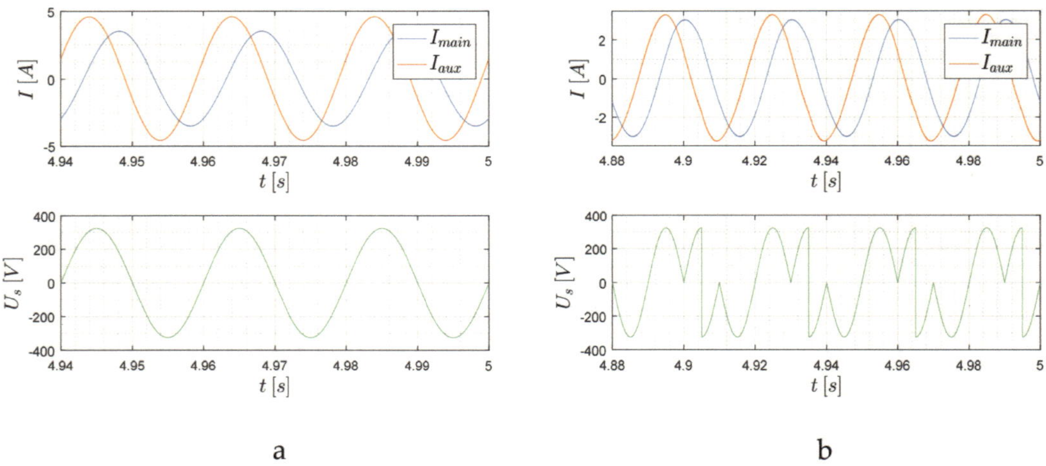

Figure 22. Main- and auxiliary-phase steady-state current (top) and SLLC MxC output voltage (bottom); (a) at 50 Hz and (b) at 33.33 Hz.

$$0 = (C_1 + C_2)D_{tc}^{2} - 2C_2D_{tc} + C_2 - \frac{1}{\omega}\frac{1}{|Z_{aux}|\cos\varphi + |\omega L_2|}. \tag{39}$$

Finding the roots of the quadratic equation we get the duty cycle D_{tc} as a function of the angular frequency ω that we can use for controlling the phase shift of the auxiliary phase

$$D_{tc}(\omega) = \frac{C_2 - \sqrt{C_{aux}(\omega)\,(C_1 + C_2) - C_1\,C_2}}{C_1 + C_2}. \tag{40}$$

In a similar way, we can derive a switching capacitor for the resonant filters, L_{res} and C_{res}. From the resonant frequency equation, we can calculate the value of resonant capacitance

$$C_{res} = \frac{1}{L\,\omega_{res}^{2}}. \tag{41}$$

Again, combining Eq. (41) for resonant capacitance with Eq. (35) for the switching capacitance $C_{res} = C$ yields

$$0 = (C_1 + C_2)D_{tc}^{2} - 2C_2D_{tc} + C_2 - C_{res}(\omega), \tag{42}$$

and

$$0 = (C_1 + C_2)D_{tc}^{2} - 2C_2D_{tc} + C_2 - \frac{1}{L_{res}\,\omega_{res}^{2}}. \tag{43}$$

On solving the equation, we get similar equation for control of resonant frequency

$$D_{tc}(\omega) = \frac{C_2 - \sqrt{C_{res}(\omega)\,(C_1 + C_2) - C_1\,C_2}}{C_1 + C_2}. \tag{44}$$

Switched capacitor will provide a requested phase shift of 90°. Requested waveforms shape should be provided by an additional LC resonant filter [19].

4.5. Single-leg matrix converter combining switched capacitors and LC filter supplying IM

Accepting of measures mentioned in Sections 4.3 and 4.4, it is possible essentially to improve quality of SLMxC. Combining switched capacitor in auxiliary phase and using LC resonant filter between the center tape of SLMxC and neutral point, it will be possible to obtain demanded current waveforms with lower value of total harmonic distortion of both main and auxiliary phases.

Worked-out simulation results are given in Section 6. At first, the simulation results of SLMxC with switched capacitors and LC filter under RL load are given in **Figure 22a** for 50 Hz, **Figure 22b** for 33.33 Hz, **Figure 23a** for 25 Hz and **Figure 23b** for 10 Hz.

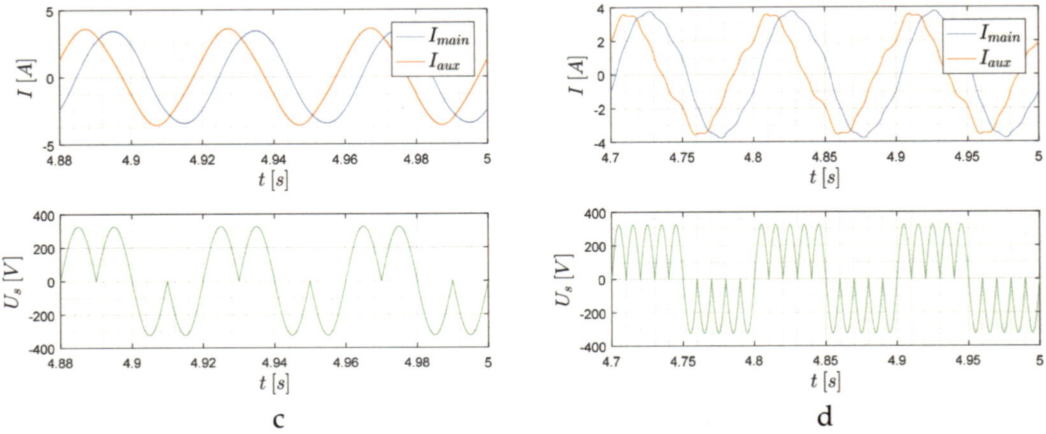

Figure 23. Main- and auxiliary-phase steady-state current (top) and SLLC MxC output voltage (bottom) at 25 Hz (c), and 10 Hz (d).

Simulation results of SLMxC with switched capacitors and LC filter under motoric IM load are shown in **Figures 24–26** in Section 6. There are shown steady-state currents and voltages of main and auxiliary phases in **Figure 24** at 50 Hz, in **Figure 25** at 33.33 Hz with PWM control, and also start-up operation in **Figure 26**.

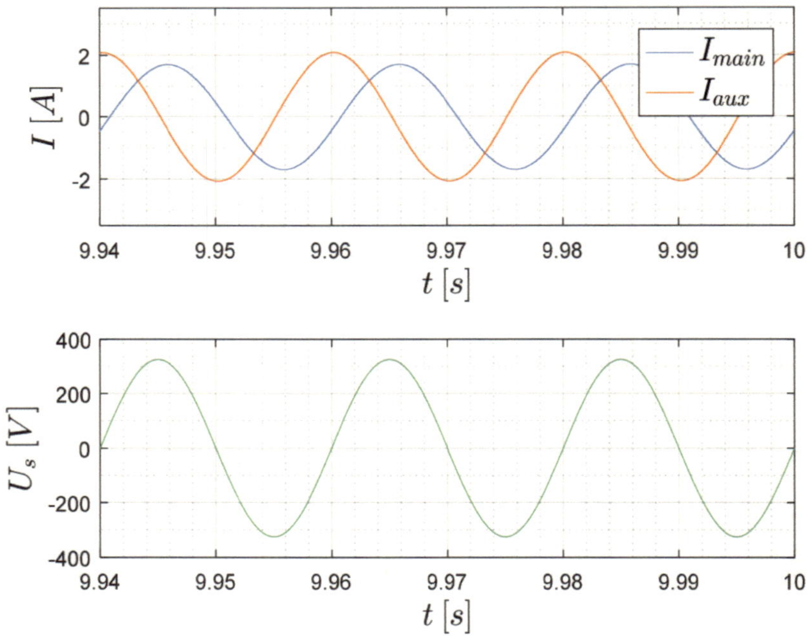

Figure 24. Main- and auxiliary-phase steady-state current of two-phase IM (top) and SLLC MxC output voltage (bottom) at 50 Hz without PWM.

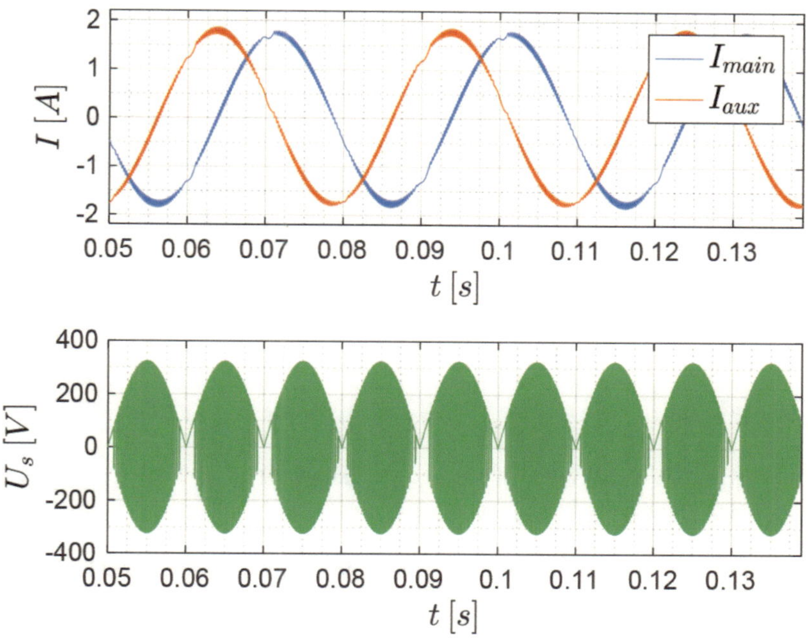

Figure 25. Main- and auxiliary-phase steady-state current of two-phase IM (top) and SLLC MxC output voltage (bottom) at 33.33 Hz with PWM.

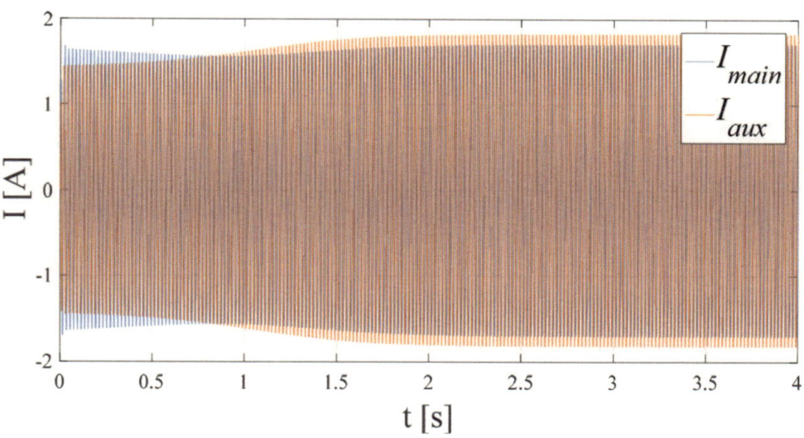

Figure 26. Main- and auxiliary-phase current of two-phase IM supplied by SLLC MxC during start-up.

5. Current controlled PWM for single-leg topologies

5.1. Current control of single-phase induction motor fed by single-leg VSI voltage source inverter

Simulation results, **Figure 27**, were worked-out without LC filter and switched capacitor [15].

Parameters used for the simulation with an induction machine:

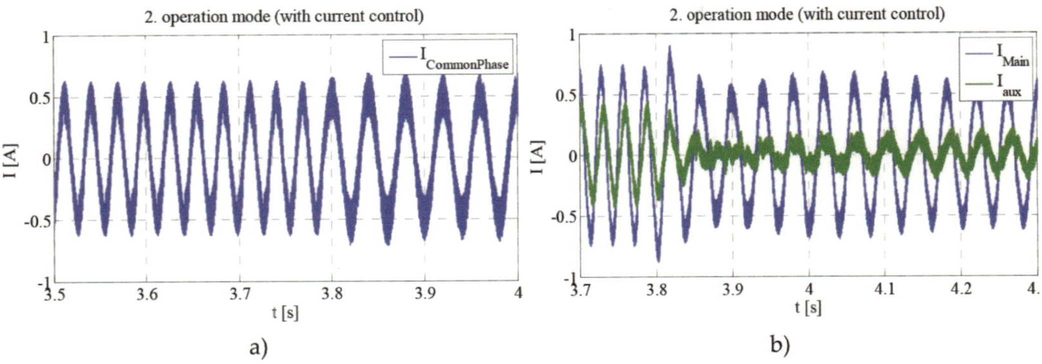

Figure 27. Currents controlled by hysteresis control; (a) in common phase and (b) in auxiliary phase [15].

$f = 50$ Hz; $U_{rms} = 230$ V; $P_{av} = 150$ W; $R_{main} = 58,85\ \Omega$; $R_{aux} = 66,1\ \Omega$; $L_{main} = 95$ mH; $L_{aux} = 120$ mH; M= 250 mH; and $C_{aux} = 20\ \mu$F.

It can be seen that the waveforms of currents of main and auxiliary phases are not shaped sufficiently. Further improving would be possible using mentioned measures, i.e. LC filter and switched capacitor.

5.2. Current control of single-phase induction motor fed by basic single-leg MxC

Basic single-leg MxC schematic is given in **Figure 11a** and **b**. As mentioned in Section 4.2, there is a problem regarding to an electromagnetic torque generated by two-phase IM motor due to very high THD of the main and auxiliary voltages and current. Therefore, neither the operation at 50 Hz nor at reduced frequency (33.33 or 25 Hz), under nominal torque and during start-up, could be provided successfully.

Parameters used for the simulation with the induction machine (the same as above):

$f = 50$ Hz; $U_{rms} = 230$ V; $P_{av} = 150$ W; $R_{main} = 58,85\ \Omega$; $R_{aux} = 66,1\ \Omega$; $L_{main} = 95$ mH; $L_{aux} = 120$ mH; M= 250 mH; and $C_{aux} = 20\ \mu$F.

Simulation results without LC filter and switched capacitor are given in **Figure 28**.

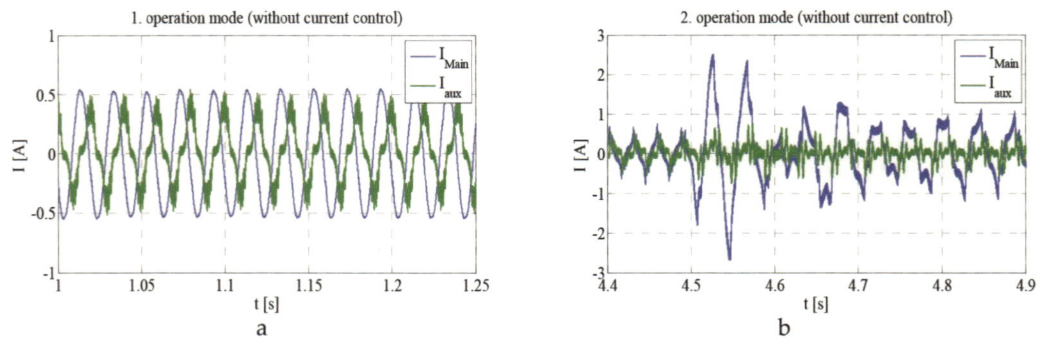

Figure 28. Stator currents of a single-phase induction motor fed by single-leg MxC; (a) in full-speed operation, and (b) reduced speed operation without LC filter and switched capacitor [17].

http://dx.doi.org/10.5772/67743

Currents are also highly deformed in both full-speed and reduced-speed regimes. In 4.5 seconds, it has changed the operation mode from full speed into reduced speed; speed of motor is proportional to the frequency of stator voltage of 25 Hz. Moreover, the start-up of the IM is not being successful. So, the main problem of single-leg matrix converter is high distortion of auxiliary-phase voltage and currents.

The basic principle of used current controlled PWM feedback loop is given in **Figure 29**.

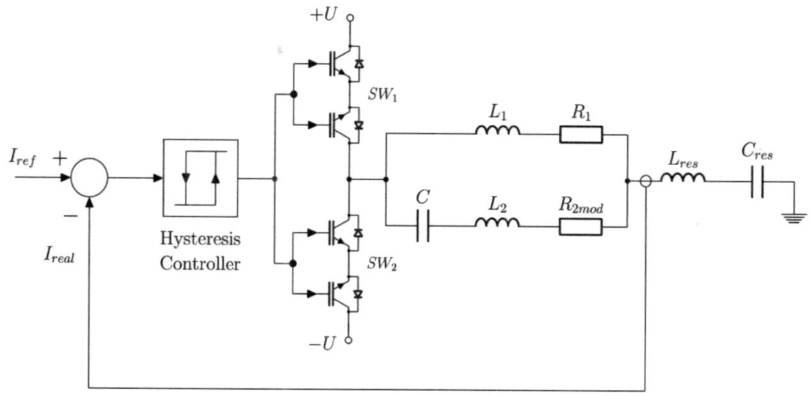

Figure 29. Principle of used CC-PWM feedback loop of single-leg MxC.

The resonant parts, L_{res}, C_{res}, have not been used at simulation of basic single-leg MxC in **Figure 28**.

5.3. Current control of single-phase induction motor fed by single-leg MxC using switched capacitor, LC filter and PWM control

Finally, combined solution with LC additional circuits and the current controlled PWM (hysteresis CC PWM) with current feedback closed loop is the best one. As mentioned in Section 4.2, simulation results of SLMxC with switched capacitors and LC filter under motoric IM load with PWM control are shown in figures in Section 6. There are shown steady-state currents and voltages of main and auxiliary phases in **Figure 25** at 33.33 Hz

Using this solution, the total harmonic distortion of main- and auxiliary-phase currents will be smaller than usually requested value of 5%.

Except the phase angle control, the amplitude of the phase currents must be controlled to optimal FOC operation of the induction machine under different load conditions.

6. Reached simulation results and discussion

All simulations of SLMxC with RL load were worked out in LT Spice environment. All simulations of SLMxC with motoric IM load were done in Matlab/Simulink programming environment.

Main- and auxiliary-phase steady-state currents and SLLC MxC output voltages have been worked-out at different frequencies 50/33.33/25/10 Hz.

The simulated phase currents under the RL load show that by control of switched capacitor the exact value of demanded capacitance is reached. Then, the phase angle between the mentioned current is also equal 90° as in **Figure 17**. From the **Figures 22–25**, it is obvious that this condition is satisfied.

Simulation of steady-state and start-up operation of two-phase IM is given in **Figures 24–26**. The simulation was done both with switched capacitor and LC resonant filter (except 50 Hz). Parameters are used for simulation with the induction machine:

$f = 50$ Hz; $U_{rms} = 230$ V; $P_{av} = 150$ W; $R_{main} = 58,85\ \Omega$; $R_{aux} = 66,1\ \Omega$; $L_{main} = 95$ mH; $L_{aux} = 120$ mH; M= 250 mH, $C_{50} = 20\ \mu F$; $L_{res} = 274$ mH, $C_{res} = 37\ \mu F$; and $q = 1$.

The simulated phase currents of the two-phase induction machine show that by control of switched capacitor reaches exact value of capacitance which in case of that the phase angle between the mentioned current is also equal 90°. The effect of the filter in a common phase will result in nearly the same magnitude of the IM currents during start-up; however, the time during start-up is rather longer.

7. Conclusion

The chapter brings analysis, modeling, and computer simulation of two-phase inverters focused on minimum switching devices. There are described two main types of switching devices: the single-leg VSI inverter partially known from a literature and single-leg MxC matrix converter as a new one. Since one-leg matrix converter type features a non-harmonic current waveform, the main emphasis is laid on the enhancement to their shapes. Because the use of classical PWM technique is restricted by insufficiency of voltage under basic frequency operation, it is necessary to use an additional hardware LC resonant circuit. After realization of above measures with LC filter, the total harmonic distortion of main- and auxiliary-phase currents will be much better: about 12% at 50 Hz and circa 9% at 33.33 Hz with LC circuits, see **Figure 24** in the text. By suitable design of LC elements, it is possible to reach the best value of the main- and auxiliary-phase current THDs (<5%) but the size of the LC elements will be high. Using that solution the total harmonic distortion of main- and auxiliary-phase currents will be smaller than usually requested value 5%. Analysis and worked-out simulation experiment results under RL load have shown that use of the LC filter can significantly improve the harmonic of the current waveform in both main- and auxiliary-phase windings. It should be also noticed that the simple LC resonant tank is always tuned to single frequency only and therefore the right operation of the MxC converter is also limited to this one frequency. To eliminate this disadvantage, the switched capacitor is supposed to use the capacitance that can be continuously changed and adapted to actual requirement given by an operational frequency. It is very important under field-oriented control of split-single-phase induction motor as a load for the converter. The basic topology has been completed by the LC filters that have both currents of main and auxiliary phases approximately sinusoidal waveforms. Main contribution of the paper is combined with the control of auxiliary phase advancing to be 90 degree under entire range of load operation and also pulse-width-modulation for field-oriented control.

Simulation experiments have been done using passive RL load and also split-winding single-phase IM motor. Worked-out results under RL load operation have shown very good

http://dx.doi.org/10.5772/67743

agreement with theoretical assumptions. Worked-out results under split-winding single-phase IM motoring operation are just preliminary ones because it needs accurate real motor parameters and takes longer time. Cooperation of switched capacitor single-leg LC matrix converter with split-winding single-phase IM is intended as for next work. The results reached can be served for usage and analysis of systems with two-phase ac motor drive. So, the next work is to focus on motoric load operation.

Acknowledgements

Results of this work were made with support of the Slovak Grant Agencies VEGA by the grant no. 1/0928/15 and APVV no. 0314/12. Authors also thank to the R&D operational program Centre of excellence of power electronics systems and materials for their components no. OPVaV-2008/01-SORO, ITMS 2622012003 funded by the European regional development fund (ERDF).

Nomenclature

LCL2C2	Center tapped series-parallel resonant LC filter
THD	Total harmonic distortion
VSI	Voltage source inverter
MxC	Matrix converter
PWM	Pulse-width-modulation
SVPWM	Space vector pulse-width-modulation
IM	Induction motor
DC	Direct current
MOSFET	Metal-oxide-semiconductor field-effect transistor
CC-PWM	Current controlled pulse-width-modulation
FOC	Field-oriented control
u_{aux}	Converter output voltage for auxiliary phase
u_{mxc}	Converter output voltage common for main and auxiliary phase
U_M	Voltage magnitude
f_{mxc}	Frequency of matrix converter output voltage
f_{ac}	Frequency of matrix converter input voltage
T_{mxc}	Period of matrix converter output voltage
T_{ac}	Period of matrix converter input voltage
Z_{main}	Impedance of main phase
Z_{aux}	Impedance of auxiliary phase
C_{aux}	Capacitance in auxiliary phase
C_{res}	Capacitance in resonant filter
L_{res}	Inductance in resonant filter

A_1	Fundamental harmonics amplitude
D_{tc}	Duty cycle
E	Energy
R_s	Resistance of IM stator winding
R_r	Resistance of IM rotor winding
L_s	Inductance of IM stator winding
L_r	Inductance of IM rotor winding
L_m	Mutual inductance
T_e	Electromagnetic torque
ω_m	Mechanical angular speed
pp	Number of pole pairs
N	Ratio between the effective numbers of turns in the auxiliary and the main stator windings

Author details

Branislav Dobrucky[1]*, Tomas Laskody[2] and Roman Konarik[1]

*Address all correspondence to: branislav.dobrucky@fel.uniza.sk

1 Department of Mechatronics and Electronics, Faculty of Electrical Engineering, University of Zilina, Zilina, Slovak Republic

2 BSH Drives and Pumps s.r.o., Michalovce, Slovak Republic

References

[1] Blalock T. J.: The first poly-phase system—a look back at two-phase power for AC distribution, IEEE Power and Energy Magazine, March-April 2004, p. 63, ISSN 1540-7977.

[2] Blaabjerg F., Lungeanu F., Skaug K., Tonnes M.: Two-phase induction motor drives, IEEE Transactions on Industry Applications, vol. 10, no. 4, July/August 2004, pp. 24–32.

[3] Blaabjerg F.: Evaluation of low-cost topologies for two-phase im drives in industrial application, Record of 37th IEEE IAS Annual Meeting on Industry Application, vol. 4, 2001, pp. 2358–2365, ISSN 0197-2618.

[4] Kinnares V., Charumit C.: Modulating functions of space vector PWM for three-leg VSI-fed unbalanced two-phase induction motors, IEEE Transactions on Power Electronics, vol. 24, no. 4, April 2009, pp. 1135–1139.

[5] Mhango L. M. C., Perryman R.: Analysis and simulation of a high-speed two-phase AC drive for aerospace applications, Electric Power Applications, IEE Proceedings, vol. 144, no.2, March 1997, pp. 149–157.

[6] Jang Do-Hyun.: PWM methods for two-phase inverters, IEEE Industry Applications Magazine, vol. 13, no. 2, March/April 2007, pp. 50–56.

[7] Klíma J.: Analytical model of induction motor fed from four-switch space vector PWM inverter. Time domain analysis, Acta Technica CSAV", vol. 44, 1999, pp. 393–410.

[8] Vodovozov V., Lillo N., Raud Z.: Single-phase electric drive for automotive applications, in Proc. of Int'l Symposium on Power Electronics, Electrical Drives, Automation and Motion SPEEDAM'14, Ischia, 2014, pp. 1293–1298.

[9] Schreier L., Chomat M., Klima J.: Analysis of three-phase induction machine operation under two-phase supply. In Proc. of the IEEE-ICIT 02 Conf., Bangkok (Thailand), 2002, ISBN 0-7803-7657-9, pp. 107–112.

[10] Chomat M., Lipo T.: Adjustable-speed single-phase IM drive with reduced number of switches, IEEE Transactions on Industry Applications, vol. 39, no. 3, May/June 2003, pp. 819–825.

[11] Chomat M., Lipo T. A.: Adjustable-speed drive with single-phase induction machine for HVAC applications, in Proc. IEEE PESC'01, 2001, pp. 1446–1451.

[12] Chomat M., Lipo T. A.: Adjustable-Speed single-phase IM drive with reduced number of switches, in Conf. Rec. IEEE-IAS Annu. Meeting, 2001, pp. 1800–1806.

[13] Lettenmainer A., Novotny D., Lipo T. A.: Single-phase induction motor with an electronically controlled capacitor, IEEE Transactions Industry Applications, vol. 27, no. 1, 1988, pp. 38–43.

[14] Danila A., Margineanu I., Campeanu R., Suciu C., Boian I.: The optimization of the single-two phase induction motor start-up with electronically switched capacitor, in Proc. of IEEE Int'l Conf. on Automation, Quality and Testing, Robotics, (AQTR 2008), Cluj-Napoca, vol. 3, 2008, pp. 450–453.

[15] Kaščák S., Laškody T., Praženica M., Koňarik R.: Current control contribution to a single-phase induction motor fed by single-leg voltage source inverter, Proc. of IEEE Int'l Conf. ELEKTRO, Zilina, 2016, DOI: 10.1109/ELEKTRO.2016.7512059.

[16] Laškody T., Dobrucký B, Kascak S, Praženica M.: 2-phase direct torque controlled IM drive using SVPWM with torque ripple reduction: Motoring and regenerating, Industrial Electronics (ISIE), 2014 IEEE 23rd International Symposium on, 1–4 June 2014, pp. 698–702.

[17] Dobrucký B., Laškody T., Praženica M.: A novel supply system for two-phase induction motor by single leg matrix converter. Elektronika ir Elektrotechnika, vol. 21, no. 4, 2015, ISSN 1392-1215, pp. 13–16.

[18] Zuckerberger A., Weinstock D., Alexandrovitz A.: Single-phase matrix converter, IEE Proc. Electric Power Appl., vol. 144, no. 4, Jul 1997, pp. 235–240.

[19] Koňarik R., Dobrucký B., Štefanec P.: Improved two-phase one-leg matrix converter using L-C filter. In Proc. of MEIT 2016 Int'l Conf. on Mechanical, Electronic and Information Technology, August 28–29, 2016, Phuket, Thailand, pp. TBA.

[20] Dobrucký B., Beňová M., Abdalmula M., Kaščák S.: Design analysis of LCTLC resonant inverter for two-stage 2-phase supply system, Automatika, vol. 54, no. 3, 2014, pp. 299–307, ISSN 0005-11-44.

[21] Laškody T.: Progressive control methods of two-phase motors, Ph.D. Thesis, Reg. No. 28260020163453, Faculty of Electrical Engineering, University of Zilina, August 2016.

[22] Dobrucký B., Praženica M., Kassa J.: Topology and control strategies for two-phase electronic converter drives, International Symposium on Power Electronics, Electrical Drives, Automation and Motion, SPEEDAM 2010, Pisa, pp. 1319–1324.

Chapter 2

A Reduced Switch Asymmetric Multilevel Inverter Topology Using Unipolar Pulse Width Modulation Strategies for Photovoltaic Application

Natarajan Prabaharan, Subramani Saravanan, Amalorpavaraj Rini Ann Jerin and Kaliannan Palanisamy

Additional information is available at the end of the chapter

http://dx.doi.org/10.5772/67863

Abstract

A new design of multilevel inverter configuration is proposed for reducing the component count and improving the quality of waveform in a photovoltaic system. The proposed configuration operates at the binary asymmetric condition for generating the large amount output voltage level with small amount harmonic distortion. Unipolar trapezoidal reference with triangular carriers is used for generating the desired switching pulses to generate the required output voltage level. The proposed configuration requires eight unipolar switches for generating the 31-level output voltage level with total harmonic distortion of 3.18% without using any filters. The value of %total harmonic distortion (THD) satisfies the IEEE 519 harmonic standard. Separate DC sources of proposed configuration are replaced by the array of photovoltaic panels for testing the configuration with the renewable energy source. The proposed configuration is tested with an experimental setup for proving the operation of it. Selected simulation and experimental results are shown for the verification of proposed configuration ability.

Keywords: multilevel inverter, pulse width modulation, trapezoidal waveform, reduced switch inverter

1. Introduction

The theory of multilevel inverter has been discussed over 30 years ago. The multilevel inverter (MLI) has many advantages when compared to conventional two-level inverter such as withstanding high voltage capability, lower harmonic distortion, lower switching losses, lower switching stress, and producing high quality of output voltage with better electromagnetic compatibility [1]. Due to that numerous advantages, the adoption of multilevel inverter has been tremendously expanded in the area of medium or high power and medium or high voltage application [2]. Generally, diode clamped multilevel inverter (DCMLI), flying capacitor multilevel inverter (FCMLI), and cascaded H-bridge multilevel (CHBMLI) are three remarkable traditional MLI topologies [3]. The drawback of conventional MLIs is the total number of components. The count of components is directly proportional to the number of levels. The balancing of voltage across DC bus capacitor in FCMLI and DCMLI is the difficult task. Also, the presence of clamping diodes and clamping capacitors in DCMLI and FCMLI, respectively, makes the circuit complex, costly and large size [4]. To overwhelm those drawbacks, numerous topologies for multilevel inverter have been introduced recently.

Reduced switch multilevel inverter configurations have their own advantages and disadvantages. In Refs. [5–20], the configurations require bidirectional switches for achieving the desired output voltage level. Utilizing of bidirectional switches increases, the total count of switches in those configurations, because the combination of two unipolar switches makes one bidirectional switch using the concept of emitter coupled to both switches. In Refs. [21–25], the transformers have utilized for generating the required output voltage level. The usage of a transformer in that configuration makes the system bulky, costly, less life span and requires more maintenance. The transformer is connected to the secondary side in series to achieve the required output voltage level. In Refs. [8, 10, 12, 14], configuration utilizes more diode and capacitors for generating the required output voltage level. The balancing of capacitor voltage is more important to achieve the particular level of the output voltage waveform.

In this chapter, the reduced switch configuration is proposed without any bidirectional switches and transformer. Therefore, the proposed configuration size and cost are considerably low. It requires only eight switches for generating the 31-level output voltage level with total harmonic distortion of 3.18%. Multicarrier unipolar trapezoidal reference with triangular carrier pulse width modulation technique is utilized for generating the switching pulses for the proposed configuration. The proposed configuration has a minimum number of conducting switches for generating per voltage level. Also, this configuration requires minimum power loss (switching loss + conduction loss) for a different number of levels when compared to other MLIs. Also, the proposed configuration is tested with the photovoltaic system for checking the ability of it.

The remaining section of this chapter is as follows: Section 2 describes the operation of proposed multilevel inverter configuration with an asymmetric condition. Section 3 describes with the multicarrier unipolar trapezoidal pulse width modulation with three different carriers such as phase disposition, alternative phase opposition and disposition and variable frequency. Section 4 describes with proposed multilevel inverter configuration integrated with the photovoltaic system. Section 5 describes with results and discussion of the proposed multilevel inverter. Section 6 ends with conclusion of this chapter.

2. Proposed multilevel inverter configurations

The proposed multilevel inverter configuration is the combination of power semiconductor switches and bypass diodes. **Figure 1a** shows the basic structure for the proposed MLI configuration. The bypass diode is connected in parallel with the combination of power semiconductor switch and DC voltage source. The basic structure has two different modes of operation. When the switch T_1 is turned on, the V_{dc1} voltage appears across the diode D1. Therefore, the value of output voltage is $2V_{dc}$ ($V_{dc1} + V_{dc2}$). When the switch T_1 is turned off, the bypass diode conducts to generate the V_{dc} output voltage. The higher number of levels can generate the cascading connection of proposed basic structure. Symmetric and asymmetric are the two different conditions of multilevel inverter for generating the output voltage level. Generally, symmetric multilevel inverter produces the minimum count of output voltage level when compared to an asymmetric multilevel inverter. **Figure 1b** shows the proposed multilevel inverter configuration

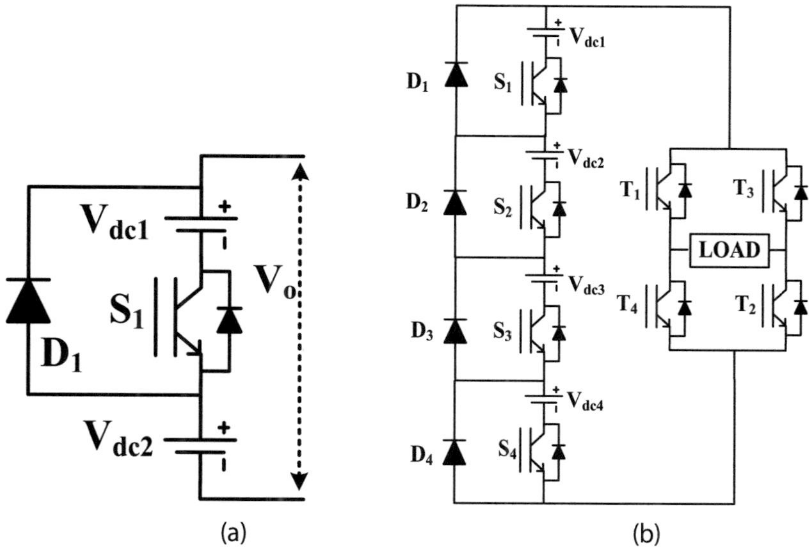

Figure 1. Proposed configuration. (a) Basic structure. (b) Proposed multilevel inverter configuration.

Parameters	Generalized formulas
Value of DC sources	2^n where $n = 0, 1, 2\ldots$
Ratio of DC sources	1: 2: 4: 8
Number of switches	$k + 4$
Number of DC source	k
Number of diodes	$k + 4$
Number of driver circuit	$k + 4$
Number of level	$2^{k+1} - 1$

Table 1. Generalized formula for the different parameters in proposed multilevel inverter.

for generating the 31-level output voltage. The DC sources are assumed as an asymmetric condition in which binary sequence is adopted. The ratio of binary sequence is 1:2:4:8. The configuration is the combination of single phase H-Bridge inverter and reduced switch configuration. The reduced switch configuration contains the set of single DC source, switch and bypass diode is connected in series. **Table 1** shows the generalized formula for the proposed multilevel inverter configuration. **Table 2** shows the switching table for the proposed topology for generating the 31-level output voltage in both positive and negative polarity.

Modes	Conducting switches and diodes	Output voltage	Modes	Conducting switches and diodes	Output Voltage
1	$S_1, D_2, D_3, D_4, T_1, T_2$	V_{dc}	1	$S_1, D_2, D_3, D_4, T_3, T_4$	$-V_{dc}$
2	$S_2, D_1, D_3, D_4, T_1, T_2$	$2V_{dc}$	2	$S_2, D_1, D_3, D_4, T_3, T_4$	$-2V_{dc}$
3	$S_1, S_2, D_3, D_4, T_1, T_2$	$3V_{dc}$	3	$S_1, S_2, D_3, D_4, T_3, T_4$	$-3V_{dc}$
4	$S_3, D_1, D_2, D_4, T_1, T_2$	$4V_{dc}$	4	$S_3, D_1, D_2, D_4, T_3, T_4$	$-4V_{dc}$
5	$S_1, S_3, D_2, D_4, T_1, T_2$	$5V_{dc}$	5	$S_1, S_3, D_2, D_4, T_3, T_4$	$-5V_{dc}$
6	$S_2, S_3, D_1, D_4, T_1, T_2$	$6V_{dc}$	6	$S_2, S_3, D_1, D_4, T_3, T_4$	$-6V_{dc}$
7	$S_1, S_2, S_3, D_4, T_1, T_2$	$7V_{dc}$	7	$S_1, S_2, S_3, D_4, T_3, T_4$	$-7V_{dc}$
8	$S_4, D_1, D_2, D_3, T_1, T_2$	$8V_{dc}$	8	$S_4, D_1, D_2, D_3, T_3, T_4$	$-8V_{dc}$
9	$S_1, S_4, D_2, D_3, T_1, T_2$	$9V_{dc}$	9	$S_1, S_4, D_2, D_3, T_3, T_4$	$-9V_{dc}$
10	$S_2, S_4, D_1, D_3, T_1, T_2$	$10V_{dc}$	10	$S_2, S_4, D_1, D_3, T_3, T_4$	$-10V_{dc}$
11	$S_1, S_2, S_4, D_3, T_1, T_2$	$11V_{dc}$	11	$S_1, S_2, S_4, D_3, T_3, T_4$	$-11V_{dc}$
12	$S_3, S_4, D_1, D_2, T_1, T_2$	$12V_{dc}$	12	$S_3, S_4, D_1, D_2, T_3, T_4$	$-12V_{dc}$
13	$S_1, S_3, S_4, D_2, T_1, T_2$	$13V_{dc}$	13	$S_1, S_3, S_4, D_2, T_3, T_4$	$-13V_{dc}$
14	$S_2, S_3, S_4, D_1, T_1, T_2$	$14V_{dc}$	14	$S_2, S_3, S_4, D_1, T_3, T_4$	$-14V_{dc}$
15	$S_1, S_2, S_3, S_4, T_1, T_2$	$15V_{dc}$	15	$S_1, S_2, S_3, S_4, T_3, T_4$	$-15V_{dc}$

Table 2. Switching table for generating 31-level output voltage in proposed configuration.

3. Switching techniques

The pulse width modulation is the most important and effective switching technique for controlling the multilevel inverter output voltage. Based on PWM technique, the output voltage can be easily converted to sinusoidal waveform by utilizing the less size of passive filters. Generally, sinusoidal pulse width modulation technique (SPWM) is utilized for generating the switching pulses to achieve the desired output voltage waveform [26–29]. In this chapter, the proposed configuration switches are triggered by using the trapezoidal reference with triangular carriers. Trapezoidal pulse width modulation technique is one of the types of advanced pulse width modulation technique. This technique provides better performance output voltage when compared with sinusoidal pulse width modulation technique which is the main advantage [28]. The combination of two slopes and one horizontal line makes the trapezoidal reference waveform. Generally, the waveform can be attained by the triangular reference waveform by limiting the magnitude or peak value of the waveform.

The angle of horizontal line of the waveform is as follows:

$$2\varnothing = (1 - \sigma)\pi \tag{1}$$

where σ is called the triangular factor. If the triangular factor is $\sigma = 1$, the waveform shape will become triangular waveform. The shape of trapezoidal waveform is purely depending on the location of slope angle (α). **Figure 2** shows the different angle of slope for the trapezoidal waveform. The harmonic content and waveform quality will differ based on the different locations of the slope angle (α). The mathematical formula for calculating the harmonic amplitude for different slope of different order is given by

$$A_n = \frac{4}{\pi}\int_0^{\pi/2} F(\theta)\sin n\theta \, d\theta \tag{2}$$

where

$$F(\theta) = \begin{cases} 1/\alpha & 0^\circ < \theta < \alpha \\ 1 & \alpha < \theta < 90^\circ \end{cases} \tag{3}$$

So, the Eq. (8) can be written as follows

$$A_n = \frac{4}{2\pi}\left[\left[\frac{-\theta k\cos(n\theta)}{n}\right]\right]_0^{\alpha} + \frac{1}{n}\int_0^{\alpha} k\cos(n\theta)\, d\theta + \frac{4}{\pi}\int_0^{\pi/2}\sin(n\theta)\, d\theta \tag{4}$$

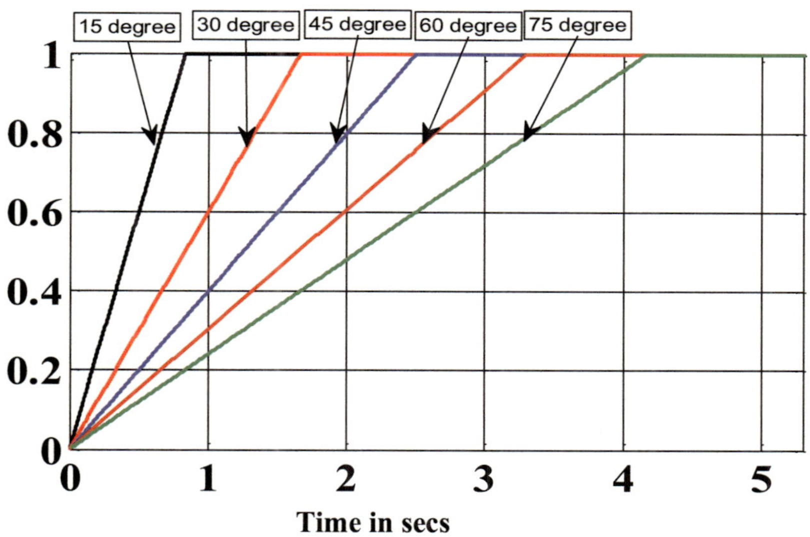

Figure 2. Different angle of slope in trapezoidal reference.

The above equation can be simplified and rewritten as follows

$$A_n = \frac{4}{n^2\pi} \times \frac{\sin(n\alpha)}{\alpha} \tag{5}$$

Figure 3 shows the harmonic content of different harmonic order for different slope angle. From that **Figure 3**, it is evident that the harmonic order value increases when the slope angle near to zero degree. If the slope angle moves towards to 90 degree, the harmonic order value decreases. In this paper, the slope of the trapezoidal reference waveform is considered as 60°. Also, in this paper, unipolar reference is considered for generating the switching pulses. In unipolar, the carriers count is reduced half of the value when compared to bipolar PWM technique, which is the main advantage of unipolar PWM method [26]. The proposed topology is tested with phase disposition (PD) carrier arrangement. Phase disposition defines that the utilization of 15 carriers is each in phase with same amplitude and frequency. The representation of unipolar trapezoidal reference with phase disposition carriers is shown in **Figure 4**.

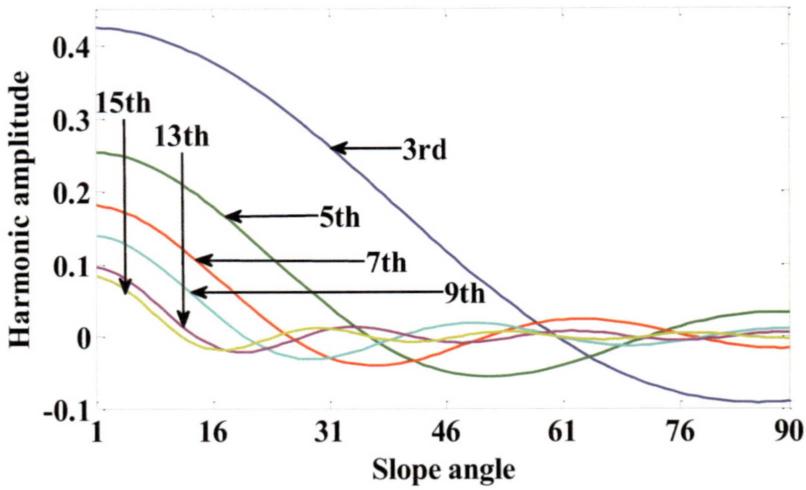

Figure 3. Harmonic content for different individual order in different slope angle of trapezoidal reference.

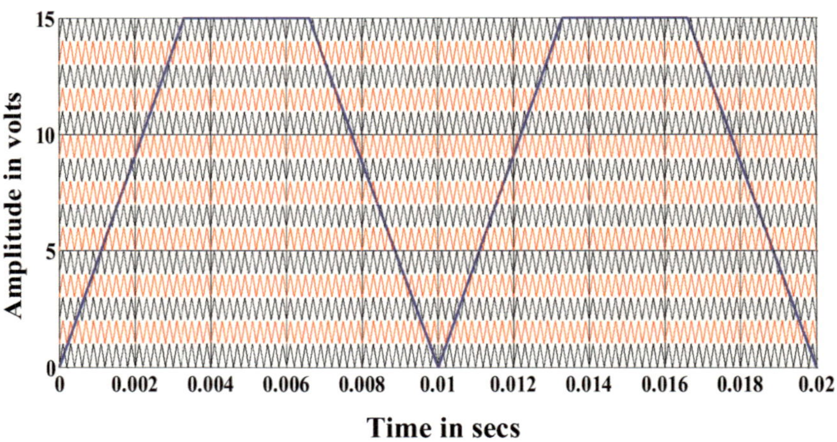

Figure 4. Unipolar trapezoidal reference with PD carrier arrangement.

4. Proposed MLI integrated with photovoltaic system

The proposed multilevel inverter requires four separate DC sources for generating the 31-level output voltage. So, the separate DC sources are replaced by the photovoltaic panel or array of photovoltaic panel depends on the input value of proposed multilevel inverter configuration. In this work, 80 W photovoltaic panel is considered. **Figure 5** shows that the proposed multilevel

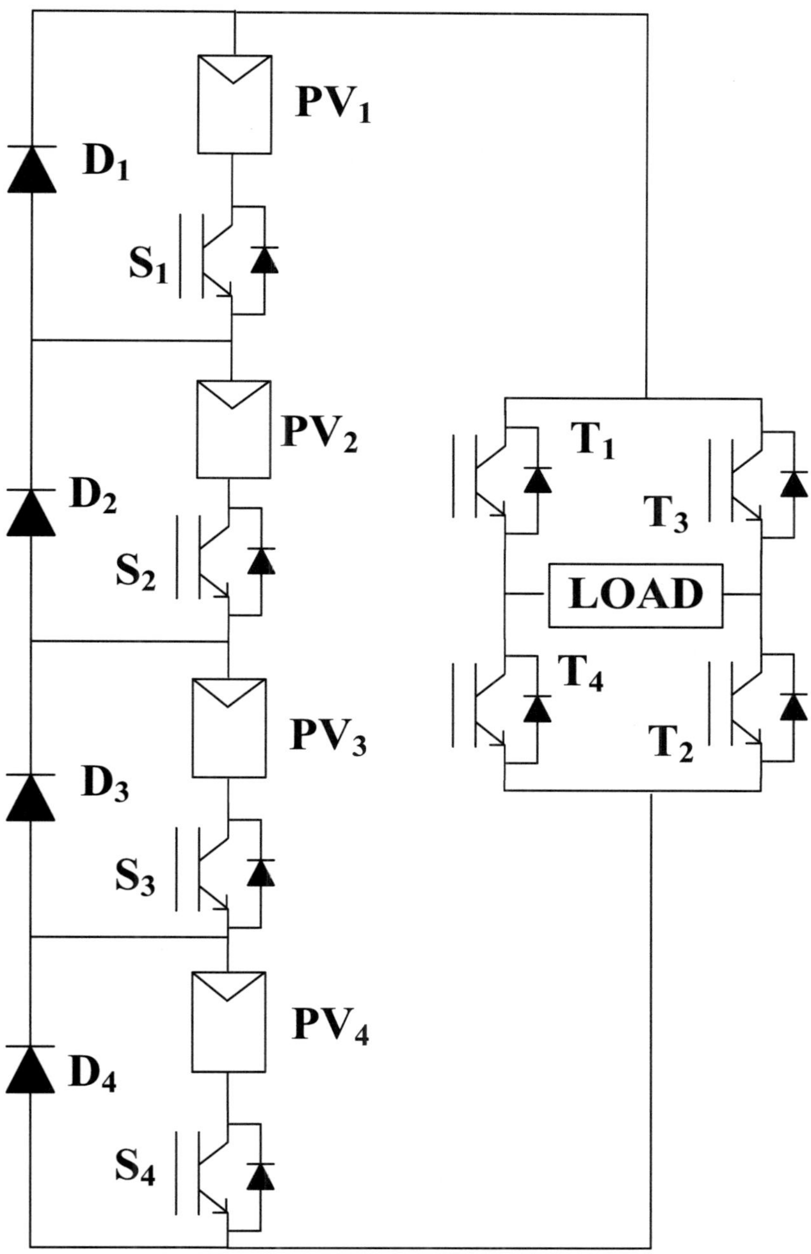

Figure 5. Proposed configuration integrated with photovoltaic system.

inverter is integrated with the photovoltaic system. For a conventional type, the integration of MLI with photovoltaic system requires separate solar panel with separate boost converter with MPPT technique for achieving the required output voltage level. Therefore, the system becomes complexity, and an initial cost of the system is too high. To overcome this drawback, this chapter proposed that the array of PV panels is connected in the series manner to achieve the required output voltage. The array of PV panels is directly connected to the proposed MLI for replacing the each DC source of it. In future, the same study is further extended with the high gain multi-output converters with appropriate MPPT techniques for reducing the count of PV array panels. The first DC source of proposed MLI is replaced by a single photovoltaic panel. The second DC source of proposed MLI is replaced by the series connection of two 80 W photovoltaic panel. The third and fourth DC sources are placed by the series connection of four panels and eight panels, respectively. **Figure 6** shows the single diode model equivalent circuit for the photovoltaic cell. **Figures 7** and **8** show the I-V and P-V characteristics of the PV model. The generalized formula for photovoltaic panel is modeled as follows [30, 31]

$$I = I_{PV} - I_o\left[exp\left(\frac{V + R_s I}{aV_t}\right) - 1\right] - \frac{V + R_s I}{R_p}\ ;\ V_t = \frac{V_s kT}{q} \tag{6}$$

$$I_{PV} = (I_{PV,n} + K_I \Delta_t)\frac{G}{G_n} \tag{7}$$

$$I_o = \frac{I_{sc,n} + K_I \Delta_t}{exp[(V_{oc,n} + K_V \Delta_t)/aV_t] - 1} \tag{8}$$

where k, a, T and q denote Boltzmann constant ($1.3806503 \times 10^{-23}$ J/K), diode ideality constant, absolute temperature (K), and electron charge ($1.60217646 \times 10^{-19}$ C), respectively. I_{PV} and I_o denote photovoltaic current and saturation current of array, respectively. $I_{PV,n}$ represents

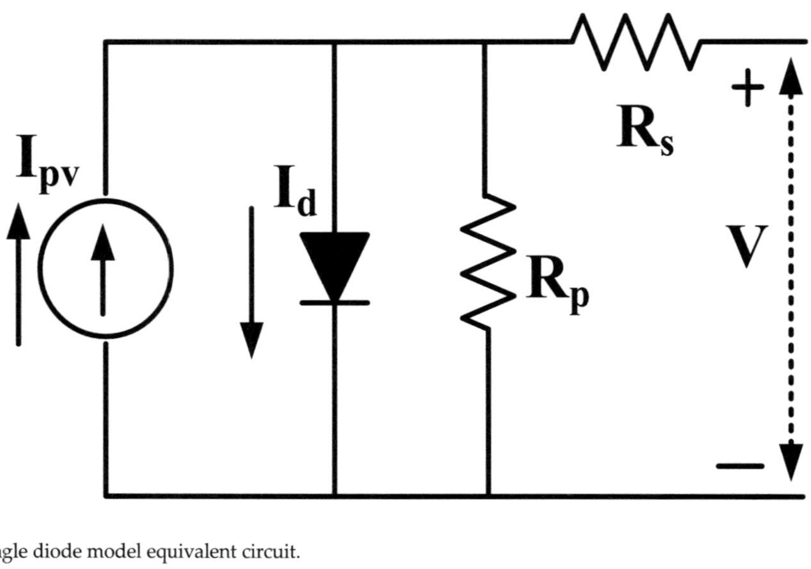

Figure 6. Single diode model equivalent circuit.

nominal PV current, and R_s and R_p denote equivalent series and parallel resistance of solar cell respectively. V_t denotes thermal voltage. N_s and N_p denote solar cells connected in series and solar cells connected in parallel, respectively. $V_{oc,n}$ and $I_{sc,n}$ indicate open circuit voltage and nominal short circuit current, respectively; K_V and K_I represent the short circuit voltage/temperature co-efficient and short circuit current/temperature co-efficient, respectively; T_n indicates nominal temperature (K); G and G_n represent irradiation (W/m^2) on the device surface and nominal irradiation and Δ_t indicates $T - T_n$, difference between actual and normal temperature. **Table 3** shows the parameters value for the 80 W photovoltaic panel.

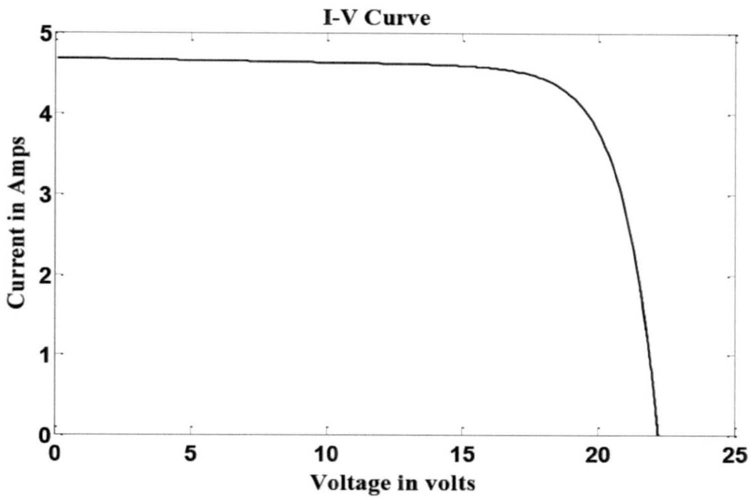

Figure 7. I-V characteristics of photovoltaic panel.

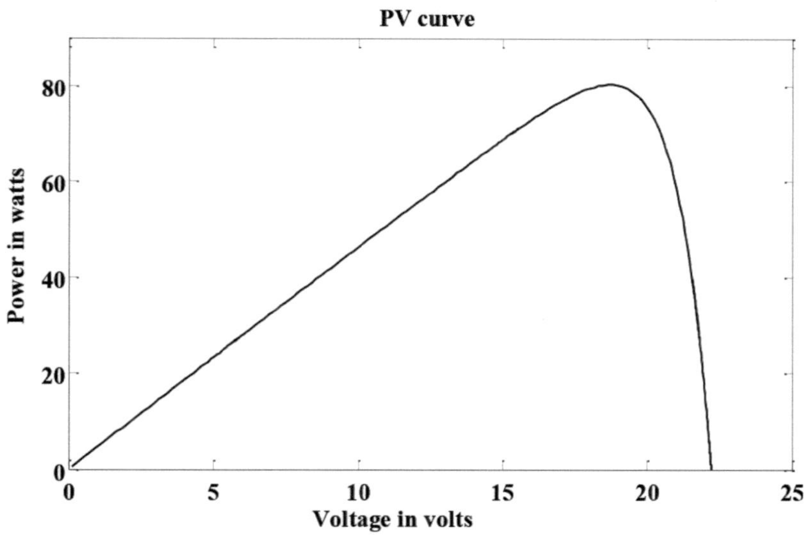

Figure 8. P-V characteristics of photovoltaic panel.

Parameters	Values
Short circuit current (I_{sc})	4.71 A
Open circuit voltage (V_{oc})	22.24 V
Maximum power point voltage (V_{mp})	18.33 V
Maximum power point current (I_{mp})	4.37 A
Maximum power (P_{mp})	80 W
Capacitors (C_1–C_4)	1500 μF

Table 3. Different parameters value for photovoltaic panel.

5. Result and discussion

The conventional CHBMLI and proposed MLI configuration are tested with MATLAB/SIMULINK for generating the 9-level and 31-level output voltage with trapezoidal pulse width modulation technique for the same number of DC source. Here, the conventional CHBMLI consists of four single H-bridge multilevel inverters which are connected in series. The conventional CHBMLI and proposed MLI configuration are tested with laboratory-based experimental set up for generating the desired output voltage using dSpace 1104 real-time controller. The unipolar PWM trapezoidal reference with triangular carriers is utilized for generating the switching pulses of the proposed multilevel inverter configuration switches in simulation and experimental step-up. Insulated Gate Bipolar Transistor (IGBT-FGA25N120) is utilized as switching devices and TLP250 as the IGBT driver for proposed topology. The main reason for selecting this PWM strategy is to produce better harmonic profile as well as to radically reduce the utilization of carrier count. **Figures 9** and **10** show the simulation 9-level output voltage, output current and their harmonics plot for output voltage, respectively, for conventional CHBMLI. **Figures 11** and **12** show the experimental result of 9-level output voltage and output current and their harmonics plot for the conventional CHBMLI. The conventional CHBMLI is

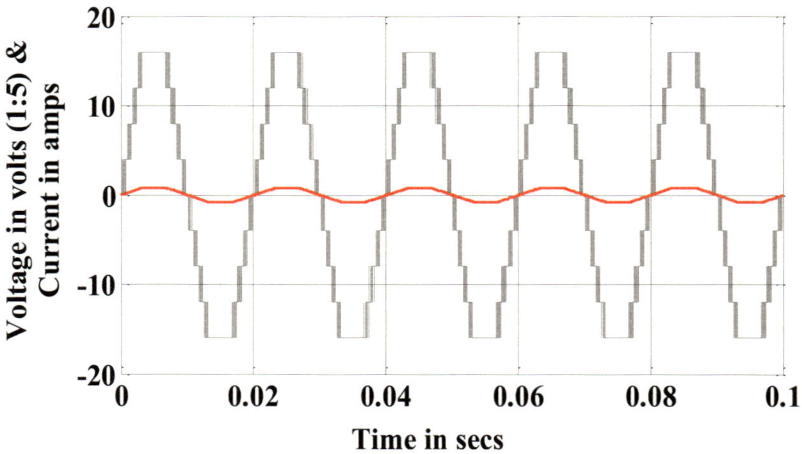

Figure 9. Simulation results for output voltage and current in conventional CHBMLI.

http://dx.doi.org/10.5772/67863

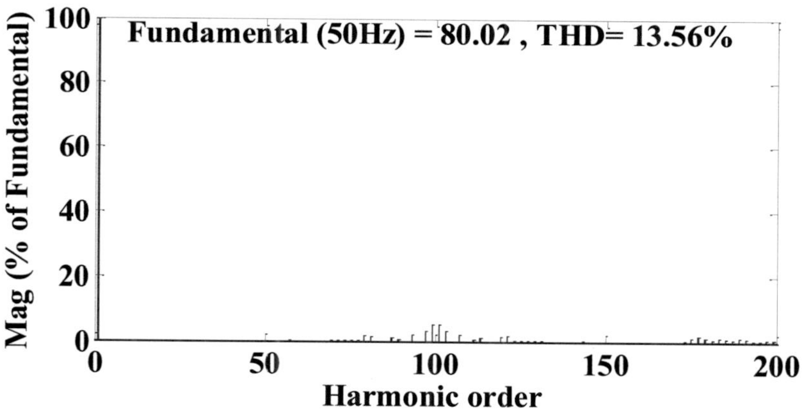

Figure 10. Harmonics plot for output voltage in conventional CHBMLI.

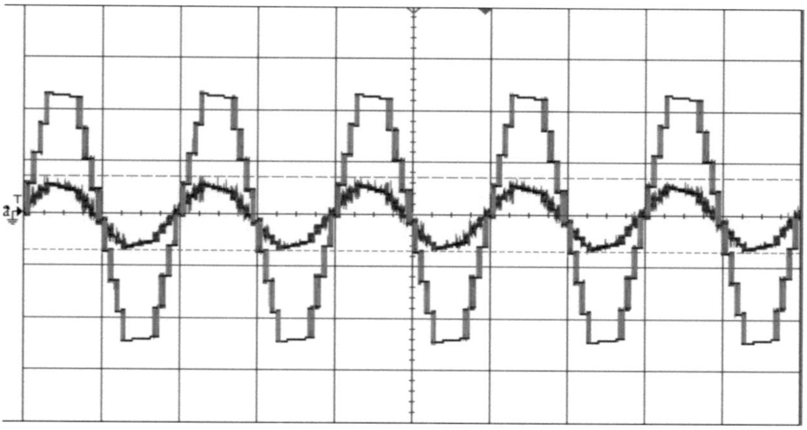

Figure 11. Experimental results for output voltage and current in conventional CHBMLI (CH1: 30 V/div, CH2: 5 A/div).

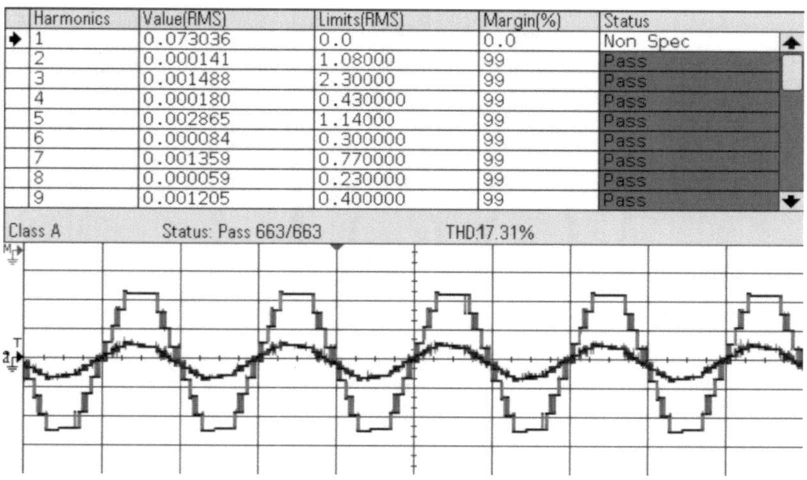

Harmonics	Value(RMS)	Limits(RMS)	Margin(%)	Status
1	0.073036	0.0	0.0	Non Spec
2	0.000141	1.08000	99	Pass
3	0.001488	2.30000	99	Pass
4	0.000180	0.430000	99	Pass
5	0.002865	1.14000	99	Pass
6	0.000084	0.300000	99	Pass
7	0.001359	0.770000	99	Pass
8	0.000059	0.230000	99	Pass
9	0.001205	0.400000	99	Pass

Figure 12. Experimental Harmonics plot for output voltage in conventional CHBMLI.

tested with the photovoltaic panels by replacing the separate DC sources. **Figures 13** and **14** show the output voltage and output current and their harmonics plot for the CHBMLI with photovoltaic systems. **Figures 15** and **16** show the simulation output voltage, output current and their harmonics plot for trapezoidal reference with PD carriers. **Figures 17** and **18** show the experimental result of output voltage and output current and their harmonics plot for the proposed topology. Also, this configuration is tested with the photovoltaic panels by replacing the separate DC sources. **Figures 19** and **20** show the output voltage and output current and their harmonics plot for the proposed configuration with photovoltaic systems. The power loss is the addition of switching losses and conduction losses. The switching losses and conduction losses can be calculated using the following formulas [18–19]

$$L_{sw} = \sum_{k=1}^{N_{sw}} \left(\sum_{i=1}^{N_{on,k}} E_{on,k,i} + \sum_{i=1}^{N_{off,k}} E_{off,k,i} \right) \quad (9)$$

$$L_c(t) = \sum_{k=1}^{N_{sw}} (L_{c,sw,k}(t) + L_{c,d,k}(t)) \quad (10)$$

Figure 21 shows that the total power loss versus a different number of levels for the proposed configuration. Also, the proposed configuration is tested with different modulation indices. **Table 4** shows the comparison table for proposed configuration with conventional MLIs in terms of many factors. From that **Table 4**, it is clearly understood that the proposed configuration requires lesser component count for generating the desired output voltage level. Also, the minimum count of conducting switches is required for generating each voltage level when compared to conventional MLIs. **Figure 22** shows the efficiency graph for the proposed configuration for different modulation indices. Therefore, the proposed configuration provides better results in terms of number of levels, efficiency and switching

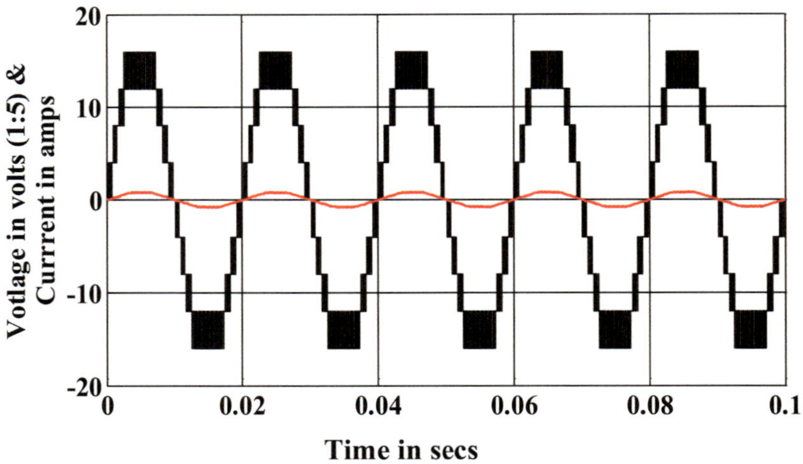

Figure 13. Output voltage and current for conventional CHBMLI integrated with PV.

http://dx.doi.org/10.5772/67863

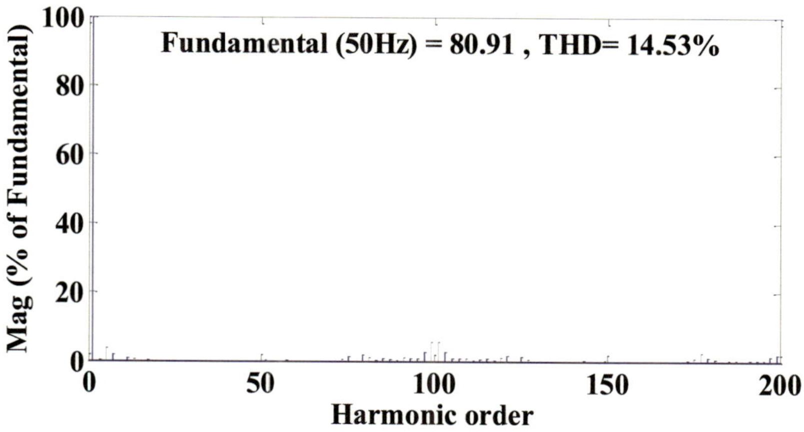

Figure 14. Harmonics plot for output voltage in conventional CHBMLI integrated with PV.

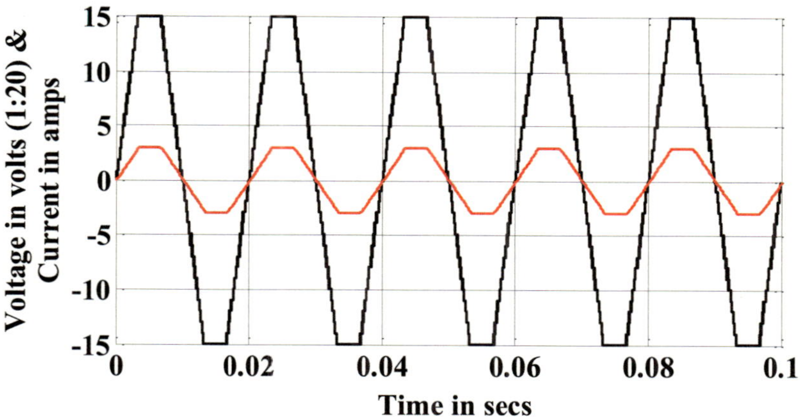

Figure 15. Simulation results for output voltage and current in proposed MLI.

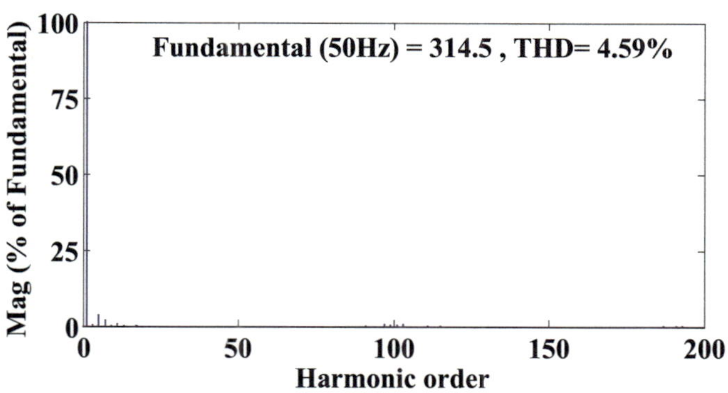

Figure 16. Harmonics plot for output voltage in proposed MLI.

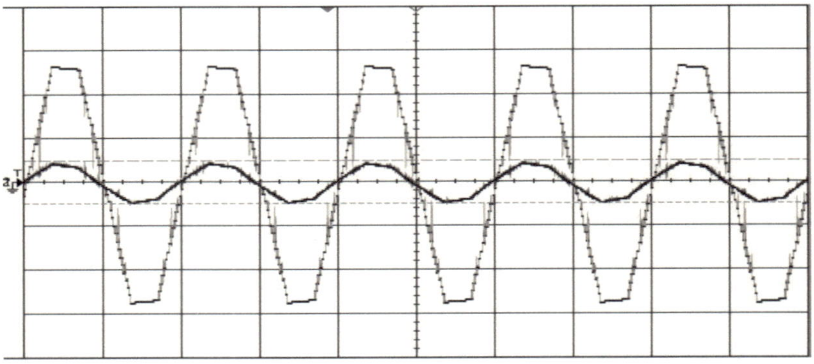

Figure 17. Experimental results for output voltage and current in proposed MLI (CH1: 100 V/div, CH2: 5 A/div).

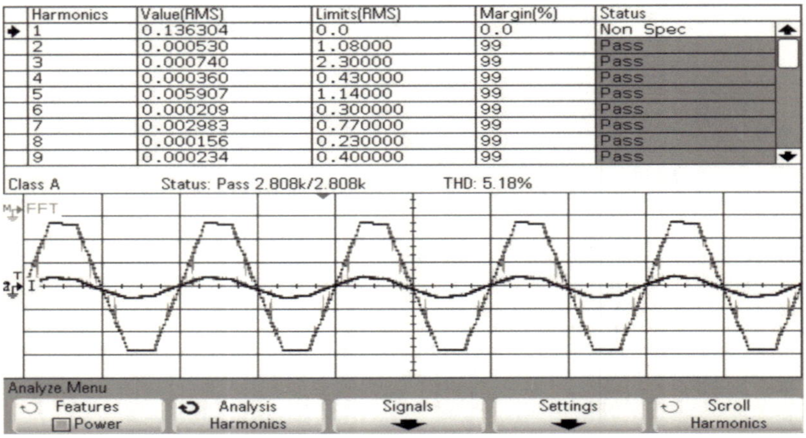

Figure 18. Experimental harmonics plot for output voltage in proposed MLI.

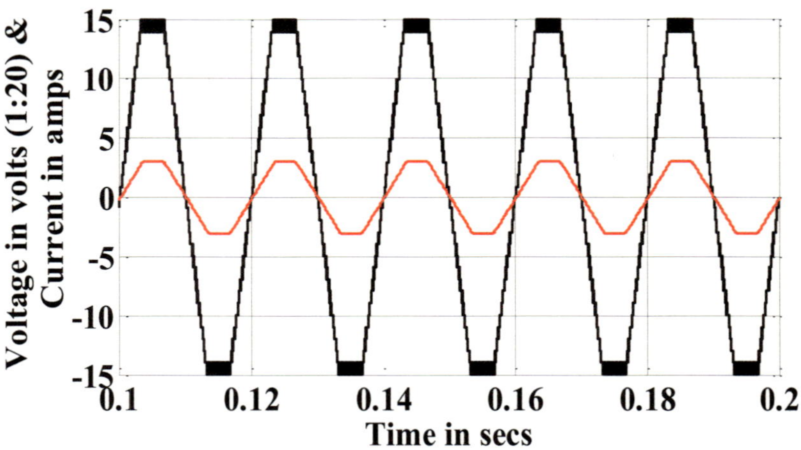

Figure 19. Output voltage and current for proposed MLI integrated with PV.

http://dx.doi.org/10.5772/67863

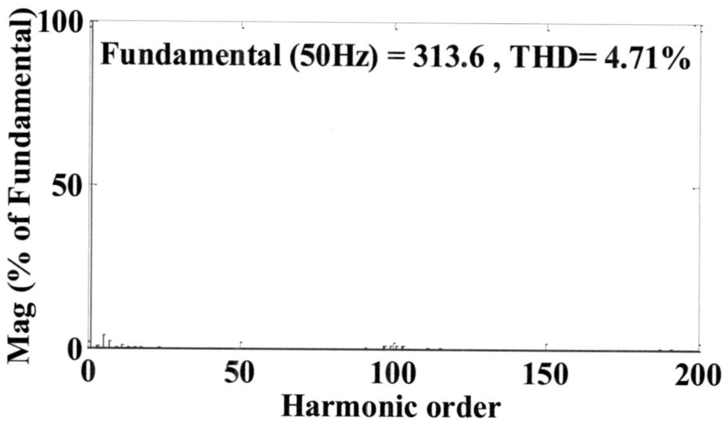

Figure 20. Harmonics plot for output voltage in proposed MLI integrated with PV.

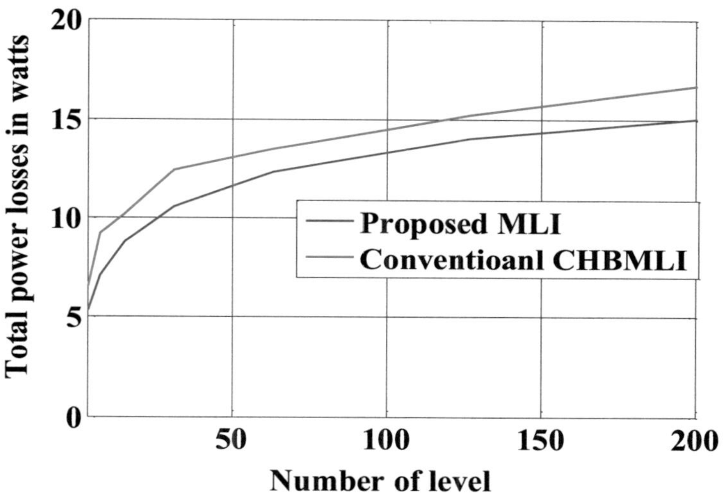

Figure 21. Power loss comparisons for proposed MLI with CHBMLI.

	DCMLI	FCMLI	CHBMLI	Proposed MLI
Number of switches	60	60	60	8
Number of DC sources	1	1	15	4
Total number of output voltage levels	31	31	31	31
Number of drive circuit	60	60	60	8
Clamping diodes	56	–	–	–
Clamping capacitors	–	28	–	–
DC bus capacitors	30	30	–	–
Conducting switches/diodes per voltage level	30	30	30	6

Table 4. Comparison table of proposed configuration with conventional MLIs for 31-level output voltage.

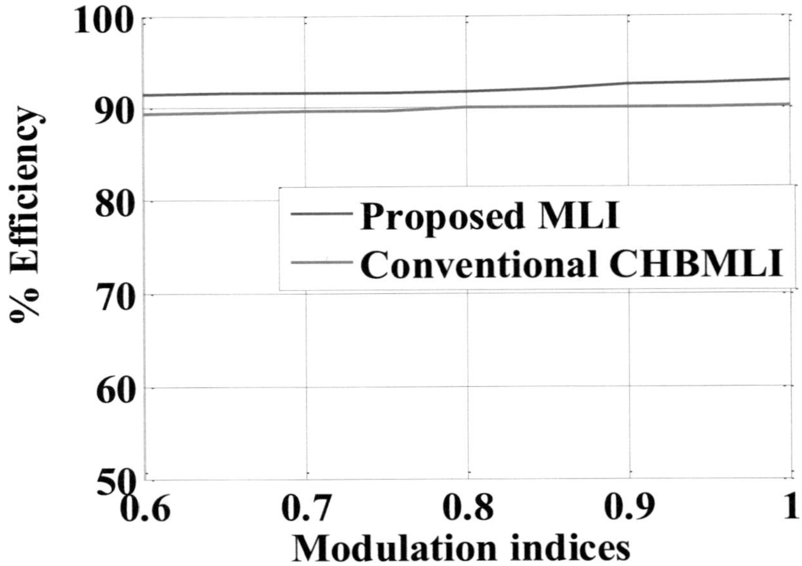

Figure 22. Efficiency comparisons for proposed MLI with CHBMLI.

Parameters	%THD without solar		%THD with solar
	Simulation	Experimental	Simulation
Conventional CHBMLI	13.56	17.31	14.53
Proposed MLI	4.59	5.18	4.71

Table 5. Comparison table of %THD for proposed MLI with conventional MLI.

losses. **Table 5** shows the %THD value for conventional CHBMLI and proposed MLI without integrated solar and with an integrated solar system. From **Table 5**, it is clearly understood that the proposed MLI with integrated solar system provides <5% THD value which satisfies IEEE519 harmonic standard.

6. Conclusion

Multilevel inverter topologies have been more popular in renewable energy application. The proposed reduced switch multilevel inverter configuration has many advantages such as reduction of switches, driver circuits and the DC source count. Also, it is operated at asymmetric condition so that it requires the minimum count of conducting switches per voltage level generation when compared with conventional MLI topologies. Therefore, the switching losses and conducting losses of this configuration are considerably low. The proposed configuration utilizes unipolar PWM strategies for improving the quality of output voltage. The operation of proposed configuration is tested with MATLAB/SIMULINK simulation, and it is verified in hardware set up using dSpace 1104 real-time controller. The proposed configuration is tested with photovoltaic panels for proving the ability of it. As a result, the proposed configuration requires lesser component count for generating higher output voltage level with lower %THD, and it is well suitable for the photovoltaic system.

Nomenclature

MLI	Multilevel inverter
PV	Photovoltaic
THD	Total harmonic distortion
CHBMLI	Cascaded H-bridge multilevel inverter
PD	Phase disposition
SPWM	Sinusoidal pulse width modulation
PWM	Pulse width modulation
I_{PV}	Photovoltaic current
I_o	Saturation current
R_s	Series resistance of solar cell
R_p	Parallel resistance of solar cell
N_s	Solar cells connected in series
N_p	Solar cells connected in parallel
G	Irradiation on the device surface
L_c	Conducting losses
$L_{c,sw}$ and $L_{c,d}$	Conducting losses of switch and conduction losses of diode
DC	Direct current
DCMLI	Diode clamped multilevel inverter
FCMLI	Flying capacitor multilevel inverter
I-V	Current-voltage
P-V	Power-voltage
FFT	Fast Fourier transform
$I_{pv,n}$	PV current
V_t	Thermal voltage
$V_{oc,n}$	Open circuit voltage
$I_{sc,n}$	Short circuit current
K_V	Short circuit voltage or temperature co-efficient
K_I	Short circuit current or temperature co-efficient
G_n	Nominal irradiation
L_{sw}	Switching losses
E_{on} and E_{off}	Turn on energy loss and turn off energy loss
N_{sw}	Number of switch

Author details

Natarajan Prabaharan*, Subramani Saravanan, Amalorpavaraj Rini Ann Jerin and Kaliannan Palanisamy

*Address all correspondence to: prabaharan.nataraj@gmail.com

School of Electrical Engineering, VIT University, Vellore, Tamil Nadu, India

References

[1] Rodríguez J, Leon JI, Kouro S, Portillo R, Prats MAM. The age of multilevel converters arrives. IEEE Industrial Electronics Magazine. 2008;**2**(2):28–39

[2] Rodriguez J, Lai JS, Peng FZ. Multilevel inverters: A survey of topologies, controls, and applications. IEEE Transactions on Industrial Electronics. 2002;**49**(4):724–738

[3] Kouro S, Malinowski M, Gopakumar K, Pou J, Franquelo LG, Bin Wu BW, Rodriguez J, Pérez MA,Leon JI. Recent advances and industrial applications of multilevel converters. IEEE Transactions on Industrial Electronics. 2010;**57**(8):2553–2580

[4] Rodriguez J, Bernet S, Steimer PK, Lizama IE. A survey on neutral-point-clamped inverters. IEEE Transactions on Industrial Electronics. 2010;**57**(7):2219–2230

[5] Samadaei E, Gholamian SA, Sheikholeslami A, Adabi J. An envelope type (E-Type) module: Asymmetric multilevel inverters with reduced components. IEEE Transactions on Industrial Electronics. 2016;**63**(11):7148–7156

[6] Hsieh C, Liang T, Chen SM, Tsai S. Design and implementation of a novel multilevel DC-AC inverter. IEEE Transactions on Industry Applications. 2016;**52**(3):2436–2443

[7] Babaei E, Hosseini SH, Gharehpetian GB, Haque MT, Sabahi M. Reduction of DC voltage sources and switches in asymmetrical multilevel converters using a novel topology. Electric Power Systems Research. 2007;**77**(8):1073–1085

[8] Islam M, Mekhilef S, Hasan M. Single phase transformer-less inverter topologies for grid-tied photovoltaic system: A review. Renewable & Sustainable Energy Reviews. 2015;**45**:69–86

[9] Babaei E. Optimal topologies for cascaded sub-multilevel converters. Journal of Power Electronics. 2010;**10**(3):251–261

[10] Babaei E, FarhadiKangarlu M. Cross-switched multilevel inverter: An innovative topology. IET Power Electronics. 2013;**6**(4):642–651

[11] Babaei E, Kangarlu MF, Sabahi M, Pahlavani MRA. Cascaded multilevel inverter using sub-multilevel cells. Electric Power Systems Research. 2013;**96**:101–110

[12] Babaei E. A cascade multilevel converter topology with reduced number of switches. IEEE Transactions on Power Electronics. 2008;**23**(6):2657–2664

[13] ShalchiAlishah R, Nazarpour D, Hosseini SH, Sabahi M. Novel multilevel inverter topologies for medium and high-voltage applications with lower values of blocked voltage by switches. IET Power Electronics. 2014;**7**(12):3062–3071

[14] Banaei MR, Salary E. New multilevel inverter with reduction of switches and gate driver. Energy Conversion and Management. 2011;**52**:1129–1136

[15] Trabelsi M, Bayhan S, Ghazi KA, Abu-Rub H, Ben-Brahim L. Finite-control-set model predictive control for grid connected packed-U-cells multilevel inverter. IEEE Transactions on Industrial Electronics. 2016;**63**(11):7286–7295

[16] Fuentes CD, Rojas CA, Renaudineau H, Kouro S, Perez MA, Thierry M. Experimental validation of a single DC bus cascaded H-bridge multilevel inverter for multistring photovoltaic systems. IEEE Transactions on Industrial Electronics. 2017; **64**(2):930–934

[17] Karasani RR, Borghate VB, Meshram PM, Suryawanshi HM. Sabyasachi S. A three-phase hybrid cascaded modular multilevel inverter for renewable energy environment. IEEE Transactions on Power Electronics. 2017;**32**(2):1070–1087

[18] Sonti V, Jain S, Bhattacharya S. Analysis of the modulation strategy for the minimization of the leakage current in the PV grid-connected cascaded multilevel inverter. IEEE Transactions on Power Electronics. 2017;**32**(2):1156–1169

[19] Lee SS, Chu B, Idris NRN, Goh HH, Heng YE. Switched-battery boost-multilevel inverter with GA optimized SHEPWM for standalone application. IEEE Transactions on Industrial Electronics. 2016;**63**(4):2133–2142

[20] Nademi H, Das A, Burgos R, Norum LE. A new circuit performance of modular multilevel inverter suitable for photovoltaic conversion plants. IEEE Journal of Emerging and Selected Topics in Power Electron. 2016;**4**(2):393–404

[21] Gandomi AA, Saeidabadi S, Hosseini SH, Babaei E, Sabahi M. Transformer-based inverter with reduced number of switches for renewable energy applications. IET Power Electronics. 2015;**8**(10):1875–1884

[22] Wang L, Zhang D, Wang Y, Wu B, Athab HS. Power and voltage balance control of a novel three-phase solid state transformer using multilevel cascaded H-bridge inverters for microgrid applications. IEEE Transactions on Power Electronics. 2016;**31**(4):3289–3301

[23] Panda AK, Suresh Y. Research on cascade multilevel inverter with single DC source by using three-phase transformers. Electric Power Systems Research. 2012;**40**:9–20

[24] Suresh Y, Panda AK. Investigation on hybrid cascaded multilevel inverter with reduced DC sources. Renewable & Sustainable Energy Reviews. 2013;**26**:49–59

[25] Veenstra M, Rufer A. Control of a hybrid asymmetric multilevel inverter for competitive medium-voltage industrial drives. IEEE Transactions on Industrial Applications. 2005;**41** (2):655–664

[26] Prabaharan N, Palanisamy K. Analysis and integration of multilevel inverter configuration with boost converters in a photovoltaic system. Energy Conversion and Management. 2016;**28**(C):327–342

[27] Prabaharan N, Palanisamy K. Comparative analysis of symmetric and asymmetric reduced switch MLI topologies using unipolar pulse width modulation strategies. IET Power Electronics. 2016;**9**(15):2808–2823

[28] Prabaharan N, Palanisamy K. Investigation of single phase reduced switch count asymmetric multilevel inverter using advanced pulse width modulation technique. International Journal of Renewable Energy Research. 2015;**5**(3):879–890

[29] Nami A, Zare F, Ghosh A, Blaabjerg F. A hybrid cascade converter topology with series-connected symmetrical and asymmetrical diode-clamped h-bridge cells. IEEE Transactions on Power Electronics. 2011;**26**(1):51–65

[30] Prabaharan N, Palanisamy K. A single phase grid connected hybrid multilevel inverter for interfacing photo-voltaic system. Energy Procedia 2016;103:250–255

[31] Prabaharan N, Palanisamy K. Modeling and analysis of a quasi-linear multilevel inverter for photovoltaic application. Energy Procedia. 2016;**103**:256–261

Chapter 3

Cascaded H-Bridge Converters Based on Current-Source Inverters: Analysis, Design, and Application on AC Drives

Pedro Eduardo Melín Coloma,
José Rubén Espinoza Castro,
Carlos Rodrigo Baier Fuentes and
Jaime Addin Rohten Carrasco

Additional information is available at the end of the chapter

http://dx.doi.org/10.5772/intechopen.68525

Abstract

This chapter reviews the cascaded H-bridge (CHB) based on current-source inverter (CSI) topology. First, the description of power topology is presented from the point of view of the current-source single-phase inverter and it connection in series with others inverters. Then, modulation of the single-phase inverter is studied, including the use of multi-level modulation techniques and their use in the proposed power topology are reviewed and simulated. Next, key design guidelines of the output capacitor and the DC inductor are reviewed. Finally, an application example for AC drives simulated in PSIM is presented. From the study, it can be concluded that the main advantage of the topology is the quality of both input currents and load voltage, while its main drawback is the use of a bulky DC inductor because of the use of current-source inverters and the oscillating power drained by the inverter from the DC side. In the same way of classic cascaded H-bridge topologies, the use of the proposal topology allows us to use semi-conductors and passive components with lower voltage and current rating than the voltage and current required by the load.

Keywords: single-phase current-source inverter, cascaded H-bridge converter based on current-source inverter, oscillating power compensation

1. Introduction

Limited voltage and current rating of semiconductors are the main limitations of the different static power converter topologies [1–4]. Diodes and thyristors are the power devices with the higher voltage blocking levels and conduction current levels, but diodes work with natural commutation while thyristors can only be communed to conduction condition. The above operation conditions do not allow one to control the electrical power transferred by the power converter—in diode-based topologies—or have poor power quality, increasing not only the input harmonic but also injecting unwanted reactive power—thyristor-based topologies - . On the other hand, force-commuted power semiconductors allow controlling the electrical power transferred and increase the power quality in both input and output of the power converter. The device's voltage and current ratings are lower than diodes and thyristors, and so in order to reach higher voltage and current levels, the devices connection in series or in parallel are typically used. From this kind of connection, and gating the devices in a convenient way, it is possible to increase further the current and voltage quality, allowing reducing losses and the size of the filter components.

The cascaded H-bridge (CHB) topologies are born under the next concept: to reach higher voltage levels using power valves with lower voltage rating, while a high power quality is keeping in the load and the power source [5–7]. Classical topology is based on H-bridge voltage-source inverters, where the series connection is natural because each inverter works as a controlled voltage source. Because of the series connection, each inverter can be disconnected from the whole array without this implying that the equipment should get offline, which is highly convenient when an inverter fails, increasing the reliability of the equipment. On the other hand, an array of nC cells per phase in a three-phase system allows dividing the load power on 3nC cells [8] so that the electrical stress in each cell is lower than other power topologies as three-phase inverters and their extension to multilevel topologies—as neutral point clamped, for example. A drawback of the cascaded connection is the power device losses which are mainly a function of the current level; in a cascaded connection, this current level is equal in all the power devices. On the other hand, for current-source converter, the natural multilevel connection is using inverters in a parallel connection. This allows summing the current injected by each converter, increasing the current waveform capability. A drawback is that the voltage rating in all the semiconductors is equal to half of the load voltage, while the capacitive filter voltage rating is equal to the load voltage.

Cascaded H-bridge based on current-source inverter (CSI) is an emerging power topology that uses a current-source inverter and a capacitive filter to synthetize a controlled voltage source that can be connected in series with other controlled voltage sources in order to reach higher voltage levels. It has been proposed for the first time in 2008 [9] for AC drive applications, and its study has been focused mainly in reducing the size of the DC inductor, the use of control techniques [10] and the compensation of using cells that are magnetically coupled [11–14], the control of the inverters using linear control and non-linear control [15–17], and the modulation and design of the power topology [18, 19].

This chapter study the cascaded H-bridge topology, without using any DC inductor reduction technique, focusing the study in operation of the power topology, the series connection of several current-source inverters in series, the use of multilevel modulation techniques, and how it defines the size of the capacitive filter required for each inverter. Also, the effect of the oscillating power drained by the inverter is described, including how it defines the size of the

http://dx.doi.org/10.5772/intechopen.68525

DC inductor is studied. Finally, the application in an AC drive computing the operation region and the key waveform of the power topology for both steady states and step changes in the DC current are studied.

2. Power topology

2.1. Power cell based on single-phase current-source inverter

Each power cell based on a single-phase current-source inverter fed by an isolated DC current source is shown in **Figure 1**. In the single-phase current-source inverter, each power valve requires symmetric blocking capabilities in order to block the AC voltages which have positives or negative values. For the abovementioned requirement, power valves can be implemented using gate turn-off thyristor (GTO) with insulated-gate bipolar transistor (IGBT) with reverse blocking capability or an IGBT with a diode in series, in order to get the reverse blocking capability. Also, new semiconductor technologies such as wide bandgap semiconductors can be used, allowing increases in the switching frequency of the power converter.

In order to simplify the power cell analysis, let's assume that we can use an ideal DC current source. This DC current source fed the single-phase inverter and, jointly, they injected a pulse width modulated current to the capacitor C_o and the load Z_L. If the modulation functions s_i of the CSI are given by

$$s_i = s_1 s_2 - s_3 s_4, \tag{1}$$

then the current injected by the CSI and the voltage in the DC side are

$$i_o = i_{dc} s_i, \tag{2}$$

and

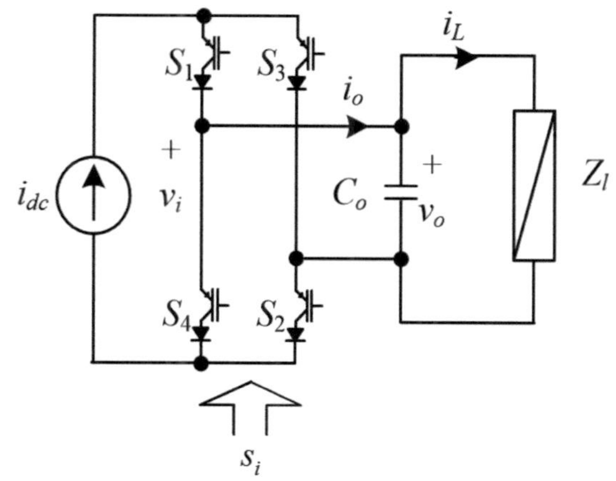

Figure 1. A power cell for CHB-CSI.

$$v_i = v_o s_i. \tag{3}$$

The modulation function s_i can be approximated to its fundamental component; then,

$$s_i \approx m_i \sin(\omega_s t + \alpha), \tag{4}$$

and the injected current and the DC voltage v_i can be defined as a function of this simplification

$$i_o = i_{dc} m_i \sin(\omega_s t + \alpha), \tag{5}$$

and

$$v_i = v_o m_i \sin(\omega_s t + \alpha). \tag{6}$$

Using the above equations, the load voltage is equal to the capacitor voltage, v_o, and both are given by

$$v_o = v_L = i_o\left(\vec{z}_L || \vec{x}_{Co}\right) \approx i_{dc} m_i \sin(\omega_s t + \alpha) Z_L. \tag{7}$$

The simplification can be made only if $\vec{z_L} << \vec{x_{Co}}$ so that all the fundamental components circulate through the load. This consideration must be included in the design requirement of the output capacitor.

2.2. Modulation and harmonic compensation on CHB-CSI

Because of the use of single-phase current-source inverter, two conditions must be avoided—(i) the electrical circuit of the DC current—typically based on an inductor—must not be open and (ii) the AC side must not be shortcircuited. The first case is because of the use of an inductor to synthetize the DC current source, and if it is open, the voltage on the power valves will theoretically become infinity; the second case is because the use of a capacitor is on the AC side. Then, if the capacitor is shortcircuited, the current on the semiconductor will be infinity. Both conditions can destroy the semiconductors used to implement the power valves.

Single-phase current-source inverter has four valid conditions (**Table 1**). Each state avoids the above conditions and allows transfer of electrical power to the load—state #1 and state #2—or disconnects the load from the DC current source—state #3 and state #4—also called zero states.

State	S1	S2	S3	S4	Io	Vi
#1	1	1	0	0	Idc	Vo
#2	0	0	1	1	−Idc	−Vo
#3	1	0	0	1	0	0
#4	0	1	1	0	0	0

Table 1. Single-phase current-source inverter states.

http://dx.doi.org/10.5772/intechopen.68525

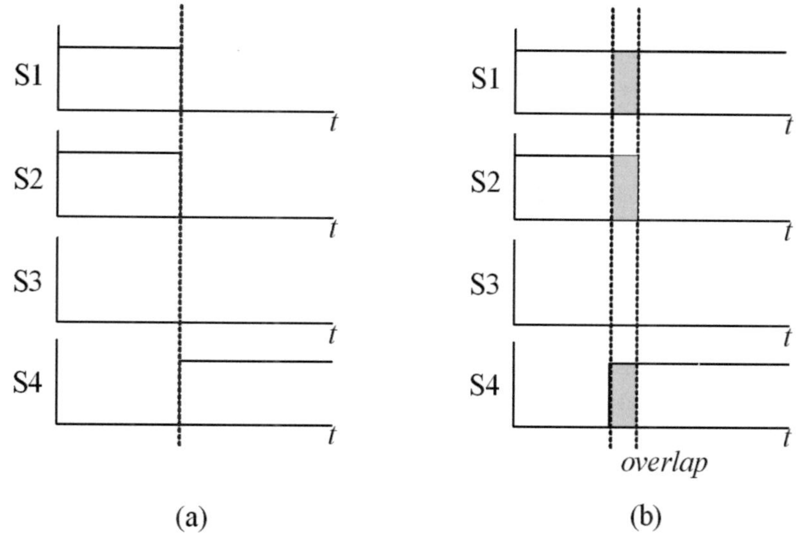

Figure 2. Transition between single-phase CSI states and overlap: (a) transitions without overlap and (b) transitions using overlap.

On the other hand, on transitions between states—**Figure 2**—it is necessary to ensure that the electrical circuit of the current source is not open. For the above, an overlap must be implemented when the inverter state is changed. The overlap should last long enough for the power valve to complete the switch. In the example, the first state is #1 and final state is #4 and in the transition between the states, the overlap is implemented.

Because of the use of a capacitive filter in each power cell, CHB-CSI topology is not a multilevel power topology but in the same way as that of multilevel topologies, the use of an appropriate modulation technique allows compensating some harmonics among inverters. In a typical current-source multilevel topology, these harmonics will be harmonic currents; in a CHB-CSI topology, the compensated harmonics will be voltage harmonics in the capacitive filter.

Sinusoidal pulse width modulation (SPWM) will be studied as an example of a modulation technique which can be used in CHB-CSI topologies. SPWM has the following advantages: it is easy to implement using both analogic circuit and digital circuit, it has the facility to modify SPWM techniques to use it in a multilevel application and the fundamental gain of the modulation technique, which in single-phase inverters, is unitary. In SPWM, a reference signal called modulator is compared with a triangular signal, also called carrier. Comparison generates a Boolean signal which is used to commutate the power valves. The inverters output signal is a pulse width modulated signal which has a wanted fundamental component and several unwanted harmonics which are the functions of the modulator frequency and the carrier fundamental frequency, so, higher carrier frequencies not only displace the unwanted harmonic to higher frequencies but also increase the commutations per period of the semiconductor devices. In multilevel topologies, the connection of the power converters in series—in the case of voltage-source converters—or parallel—in the case of current-source converters—allows to sum up the DC voltage/current levels and compensate the unwanted harmonics if they are generated and phase-shifted among them in an appropriate way. In case of phase-shifted carrier (PSC) sinusoidal pulse width modulation, the switching signals are generated comparing n_C carriers phase-shifted at $180°/n_C$, among them with a common modulator signal. An example simulated in MATLAB is shown in **Figure 3**. In the first case, the modulator signal is compared with the carrier, generating the pulse width modulation signal shown below (**Figure 3a**). Multilevel cases are **Figure 3b** and **Figure 3c** for $n_C = 2$ and $n_C = 3$, respectively, where the resulting waveform is of 5 levels for $n_C = 2$ and 7 levels for $n_C = 3$. Computing and comparing the total harmonic distortion (THD) of the three PWM signals presented, these values are 46, 25, and 14% for $n_C = 1$, $n_C = 2$, and $n_C = 3$, respectively, showing the reduction of the distortion of the resulting pulse width modulated waveform without increasing the commutation frequency. The above is valid for multilevel topologies. The effects of using PSC SPWM in a CHB-CSI topology—which is not a multilevel topology—will be analyzed in the next section.

2.3. Cascaded connection of single-phase CSI

Inverters with their isolated and controlled DC current source and their capacitive filter can be connected in a series array because each power cell is working as a controlled AC voltage source (**Figure 4**). With the above, the voltage of the array is the sum of all power cells connected to it, allowing (i) to use components with lower voltage ratings than the voltage of the application and (ii) to divide the power of the application in multiple power cells. **Figure 4** shows multiple power cells—which will be named as $3n_C$—feeding a common three-phase load. Each cell injects a controlled current to the load. Defining $\vec{i}^{\,j}_{oi,1}$ as the fundamental current of the $i = 1, 2, \ldots, n_C$ cell feeding the j = u,v,w phase and $\vec{i}^{\,j}_{L,1}$ as the fundamental component of the load current of the same j phase, then

$$\vec{i}^{\,j}_{L,1} = \vec{i}^{\,j}_{o1,1} = \vec{i}^{\,j}_{o2,1} = \ldots = \vec{i}^{\,j}_{on_C,1}, \tag{8}$$

while the load voltage is the summation of the cell output voltage, each one is given by the voltage on the capacitor filter so

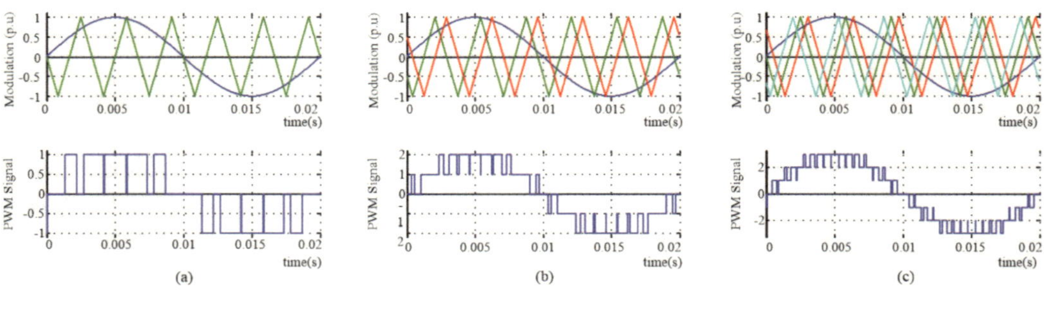

Figure 3. Sinusoidal pulse width modulation and phase-shifted carrier sinusoidal pulse width modulation with a modulation signal equal to 50 Hz; (a) modulator, carrier signals, and 3 levels of pulse modulation signal, (b) modulator, carrier signals, and 5 levels of pulse modulation signal, and (c) modulator, carrier signals, and 7 levels of pulse modulation signal.

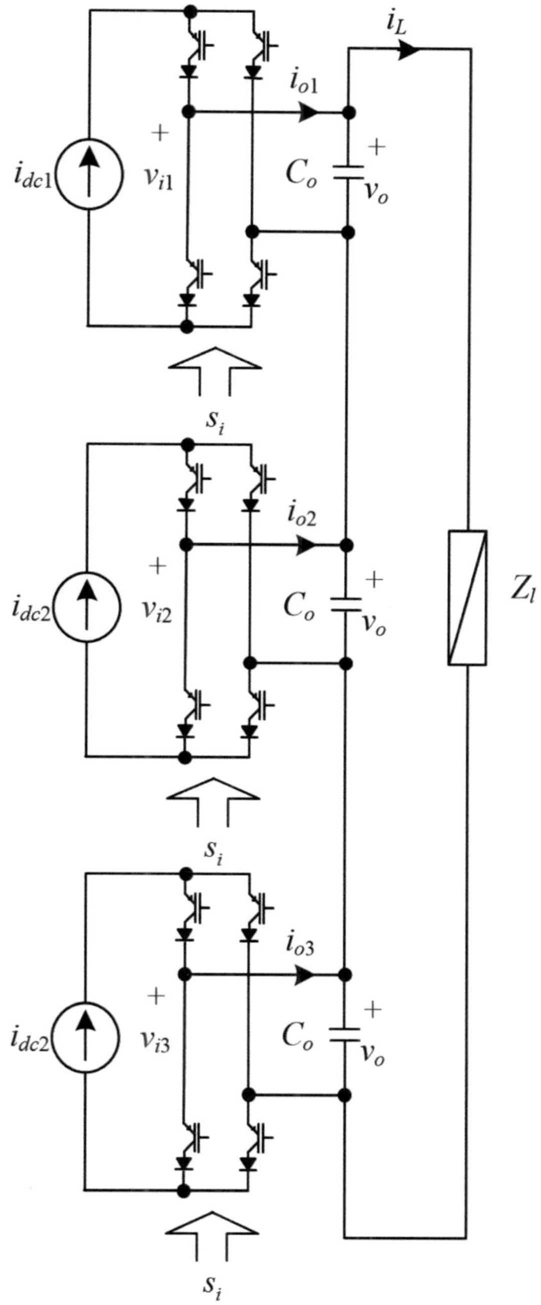

Figure 4. Three-phase applications of CSC-CHB ($n_C = 3$).

$$\vec{v}^{\,j}_{L,1} = \vec{v}_{N\,j,1} = \sum_{i=1}^{n_C} \vec{v}^{\,j}_{oi,1} = \vec{i}^{\,j}_{L,1}\vec{z}_{L,1}. \tag{9}$$

From the equations, it is clear that the connection of single-phase inverters in a cascaded array allows dividing the load voltage *in N* cells, allowing the use of a semiconductor with lower voltage rating than the required load voltage, but the current in each cell is equal to the other, increasing semiconductor losses.

An advantage of the topology is the quality of the voltage waveform at the load. Because of the use of a capacitive filter, the sum of all cell output voltages is not a multilevel voltage, but through the multilevel modulation technique that is used in each inverter connected in series, it is possible to compensate the dominant harmonic among cells. An example of the above is shown in **Figure 5**, where the topology has been simulated using PSIM in order to obtain the load voltage waveform for $n_C = 1$, $n_C = 2$, and $n_C = 3$. For $n_C= 2$ and $n_C= 3$, a multilevel modulation technique—specifically phase-shifted carried pulse width modulation—is used. For $n_C = 1$, one has THD = 28.2% which is reduced to THD = 9.6% in $n_C = 3$. The above is because dominant harmonic presented in each capacitor is phase shifted with the dominant harmonic in the other capacitor. Each capacitor voltage for N = 2 and $n_C = 3$ can be seen in **Figure 5d** and **e**, where each waveform is similar to the voltage capacitor in $n_C = 1$.

2.4. Isolated DC current source for power cells

Each power cell requires an isolated DC current source and there are several options to implement the DC current source. For example, a controlled rectifier in series with a DC

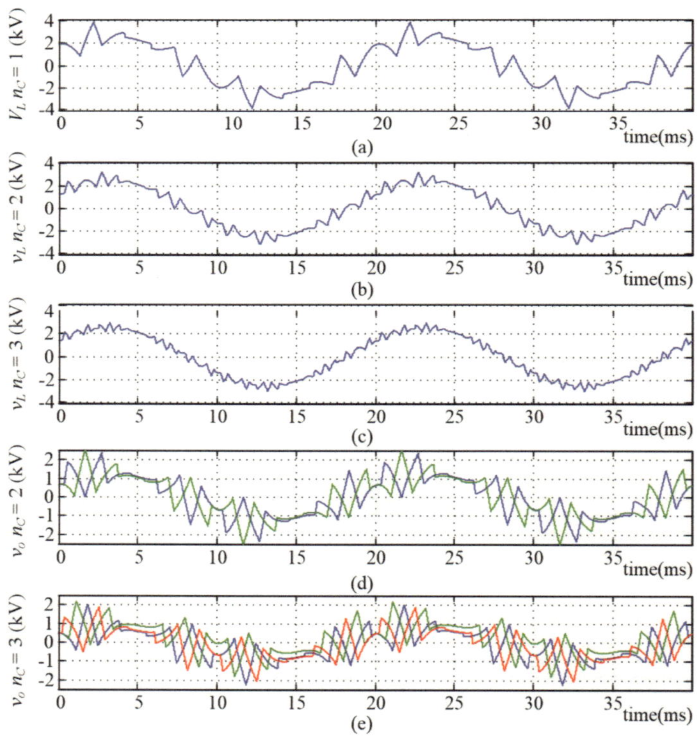

Figure 5. Voltage features for $n_C = 1$, $n_C = 2$, and $n_C= 3$, $C_o = 17\mu F$; (a) load voltage for $n_C = 1$ (THD = 28.2%), (b) load voltage for $n_C = 2$ (THD = 14.2%), (c) load voltage for $n_C = 3$ (THD = 9.6%), (d) capacitor voltage for $n_C = 2$, and (e) capacitor voltages for $n_C = 3$.

reactor can be used to get, from the viewpoint of the inverter, a controlled DC current source. In order to isolate this DC current source from other DC current sources that feed other power cells, a power transformer is required. With the above, the rectifier stage can be implemented using single-phase or three-phase controlled rectifiers based on thyristor or force-commuted semiconductors as IGBT or silicon carbide (SiC). Both cases require a DC reactor on the DC side; in thyristor rectifier cases, it can be connected directly to the secondary transformer while in force-commuted semiconductor rectifiers, an LC filter is required between the secondary transformer and the rectifier. Another option is to use a diode rectifier and a DC/DC converter on the DC stage. This case limits the power that can be transferred to the inverter stage but is a good option for non-conventional renewable power source. Also, if the source is a DC power source type, a DC/DC to regulate the DC current to the inverter stage can be used. This is the case of photovoltaic arrays and fuel cells. A third case is when the inverters are directly connected to the power grid. In this case, the DC current source can be implemented with the single-phase inverter and the DC inductor. The DC current regulation must be implemented in the inverter control scheme.

3. DC Reactor on cascaded H-bridge based on current source inverters

3.1. Oscillating power on single-phase current-source inverter

Using single-phase inverters involving the occurrence of an oscillating and continuous power, it can be described as

$$p_o = S_o\left[\cos\left(\phi_m\right) - \cos\left(2\omega_i t + 2\alpha_i + \phi_m\right)\right], \tag{10}$$

where $S_o = \frac{1}{2} Z_m I_{dc}{}^2 M_i{}^2$ is the load apparent power, M_i is the inverter modulation index—considering fixed for this case—and Z_m is the equivalent impedance of the load in parallel with the inverter output capacitor as is shown in **Figure 6**. Then, the power drained from the cell can be written in terms of the cell energy, and the charge and discharge of the DC inductor current is

$$P_{cell} = \frac{\Delta E_{cell}}{\Delta t} = \frac{L_{dc}\left[i_{dc}(t_2)^2 - i_{dc}(t_1)^2\right]}{2(t_2 - t_1)} = p_o(t_2) - p_o(t_1), \tag{11}$$

where $i_{dc}(t_1) = I_{dc}$ e$i_{dc}(t_2) = I_{dc} k_{dc}$, and a k_{dc} near to 1 means that there is no variation in the DC current level. With the above, one can write

$$4\omega_i \pi^{-1} L_{dc} I_{dc}{}^2\left[(k_{dc})^2 - 1\right] = S_o. \tag{12}$$

Then, the DC current variation, in per unit, can be written as

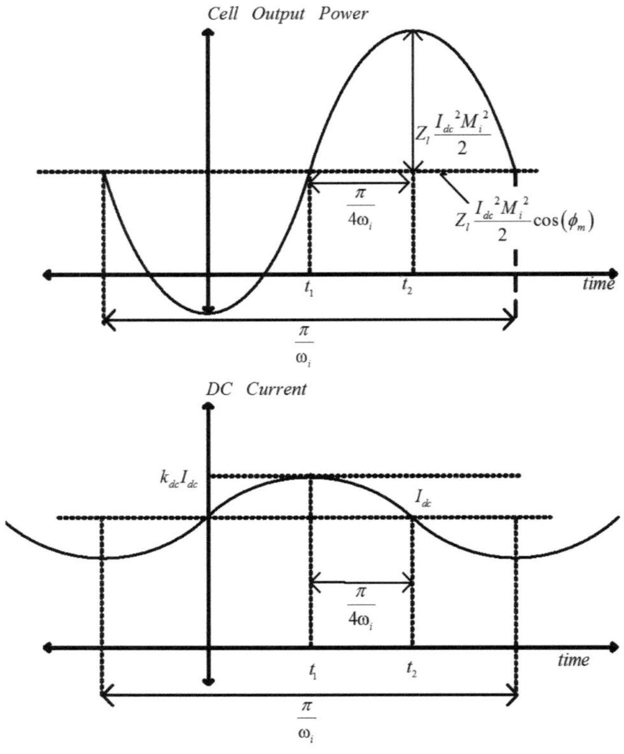

Figure 6. Oscillating power and its effects in the DC current.

$$k_{dc} = \sqrt{\frac{\pi}{8\omega_i}\frac{Z_m M_i^2}{L_{dc}} + 1}. \tag{13}$$

Considering that the power converter that feeds the DC inductor and the single-phase inverter only injects to the DC side, the continuous power drained by the inverter—corresponding to the active power delivered by the inverter to the load—and that the voltage imposed by power converter in the DC side has mainly a DC component, another element must provide the oscillating power so that it does not disturb the DC current.

The easier solution to avoid the DC current variation due to the oscillating power is to increase the size DC inductor. This increases the losses in the DC link, along with the size of the inductor and increases the cost of each cell. Other options consider the use of active compensators in the DC side, tuned passive filters, or neutral leg. Specifically, for the case of an AC drive, the active compensation adds complexity to the topology, adding semiconductors and additional accumulators to DC link, besides requiring additional controllers to manipulate the semiconductors incorporated, but it is a good option when the inverters are directly connected to the electrical grid or when the CHB-CSI topology is used in photovoltaics application. On the other hand, the passive techniques are mainly applied to cases where the frequency inverter is fixed, making a complex application for AC drives.

3.2. DC inductor design

The main objective of the DC inductor is to limit the DC current variation. Due to the oscillating power drained by the single-phase inverter and its effects on the load current and load voltage, the lower frequency that the DC inductor must limit to is the second harmonic of the inverter frequency. Then, using Eq. (11) and defining k_{dc} in Eq. (13), the DC inductor can be computed with

$$L_{dc} = \frac{\pi}{8\omega_i} \frac{Z_m M_i^2}{\left(k_{dc}^2 - 1\right)}. \tag{14}$$

4. Capacitive filter

4.1. Capacitive filter design as a function of the load voltage THD, $n_C = 1$

For $n_C = 1$, the inverter output voltage v_o is equal to the load voltage. Also, the output voltage total harmonic distortion is given by

$$THD_{Vo} = \frac{\sqrt{\sum_{i=2}^{\infty} V_{o,h}^2}}{V_{o,1}}, \tag{15}$$

where fundamental output voltage is defined by

$$V_{o,1} = I_{dc} M_i G_{ac} Z_m, \tag{16}$$

with M_i as the modulation index and

$$Z_m = |\vec{z}_M| = |\vec{x}_{Co}||\vec{z}_L|. \tag{17}$$

On the other hand, output voltage harmonics are defined by the inverter current harmonics that flow through the capacitor. Then

$$V_{o,h} = I_{dc} f_{iac,h} X_{Co}. \tag{18}$$

Hence, Eq. (15) can be written as

$$THD_{Vo} = \frac{\sqrt{\sum_{i=2}^{\infty} V_{o,h}^2}}{V_{o,1}} = \frac{X_{Co}\sqrt{\sum_{i=2}^{\infty} \left(\frac{f_{iac,h}}{h}\right)^2}}{Z_m M_i G_{ac}}. \tag{19}$$

Eq. (19) shows that the output voltage THD is a function of the filter reactance, impedance Z_m, and modulation techniques, including the modulation index. Grouping the terms defined by the modulation technique, F_{iac} can be defined as

$$F_{iac} = \frac{\sqrt{\sum_{i=2}^{\infty}\left(\frac{f_{iac,h}}{h}\right)^2}}{M_i G_{ac}}. \tag{20}$$

Considering Eq. (17) on Eq. (15) and solving for C_o, it can be found that

$$C_o = \left(\omega_i^2 L_L \pm \omega_i \sqrt{\left(\frac{THD_{Vo}^2}{F_{iac}^2}\right)\left((R_L)^2 + (\omega_i L_L)^2\right) - R_L^2}\right)^{-1}, \tag{21}$$

where two solutions for C_o can be computed due to the ± sign in the denominator.

4.2. Capacitive filter design for n_C inverters

The fundamental load voltage is the summation of each capacitor's fundamental voltage (Eq. (9)) and this voltage will not change for $n_C \neq 1$ in order to keep the performance of the array, that is, there is a symmetrical distribution of the load voltage among cell output capacitors. The voltage in each capacitor is the function on the fundamental current injected by the inverter, the load impedance, and the number of inverters connected in series, while the current harmonics generated by the inverters must flow across the capacitive filter of each cell, which will generate voltage harmonics in each capacitor and, therefore, the sum of these voltage harmonics will be in the load voltage. Then, it is possible to write Eq. (15) as

$$THD_{V_L} = \frac{X_{Co}\sqrt{\sum_{i=2}^{\infty}\left(\sum_{j=1}^{n_C}\frac{f_{iacj,h}}{h}\right)^2}}{M_i G_{ac}\sum_{j=1}^{n_C}\frac{Z_M}{n_C}} = \frac{X_{Co}}{Z_M}F_{iacM}, \tag{22}$$

where the terms by the modulation technique can be summarized in one term defined by

$$F_{iacM} = \frac{\sqrt{\sum_{i=2}^{\infty}\left(\sum_{j=1}^{n_C}\frac{f_{iacj,h}}{h}\right)^2}}{M_i G_{ac}}. \tag{23}$$

Finally, C_o can be found solving Eqs. (22) and (23) for an RL load as shown in Eq. (24). Then, with n_C inverters in a cascaded device, n_C capacitors are needed, one for each inverter, and they can be computed using

$$C_o = \left(\omega_i^2 L_L \pm \omega_i \sqrt{\left(\frac{THD_{Vo}^2}{F_{iacM}^2}\right)\left((R_L)^2 + (\omega_i L_L)^2\right) - R_L^2}\right)^{-1}. \tag{24}$$

Due to the series connection of the current-source inverters, it is not possible to sum up the current level and obtain a multilevel current waveform—multilevel current source topologies must be connected in a parallel array to sum up currents levels, but it is possible to compensate voltage harmonic among the capacitor voltage. These harmonics are generated by the current harmonic injected by the inverters and are a function of the switching function—see Eq. (22)—therefore, if a multilevel modulation technique is used with the aim of generating current harmonics that are phase shifted, the DC current level in each inverter is the same and all outputs filters have the same capacitor value; some capacitor voltage harmonics will be phase shifted and can be compensated among cells. The amplitude of the voltage harmonic in each cell is a function of the capacitor value—see Eqs. (21) and (24)—so by increasing the capacitor size, the voltage harmonic will be increased and, at the same time, the capacitor voltage rating.

As examples, values of F_{iac} and F_{iacM} are computed for PSC-SPWM using MATLAB and they are presented in **Figure 7**, considering modulation indexes $0.5 \le M_i \le 1$ and different carrier frequencies for each case. For $n_C = 2$ and $n_C = 3$, if a multilevel modulation technique is not

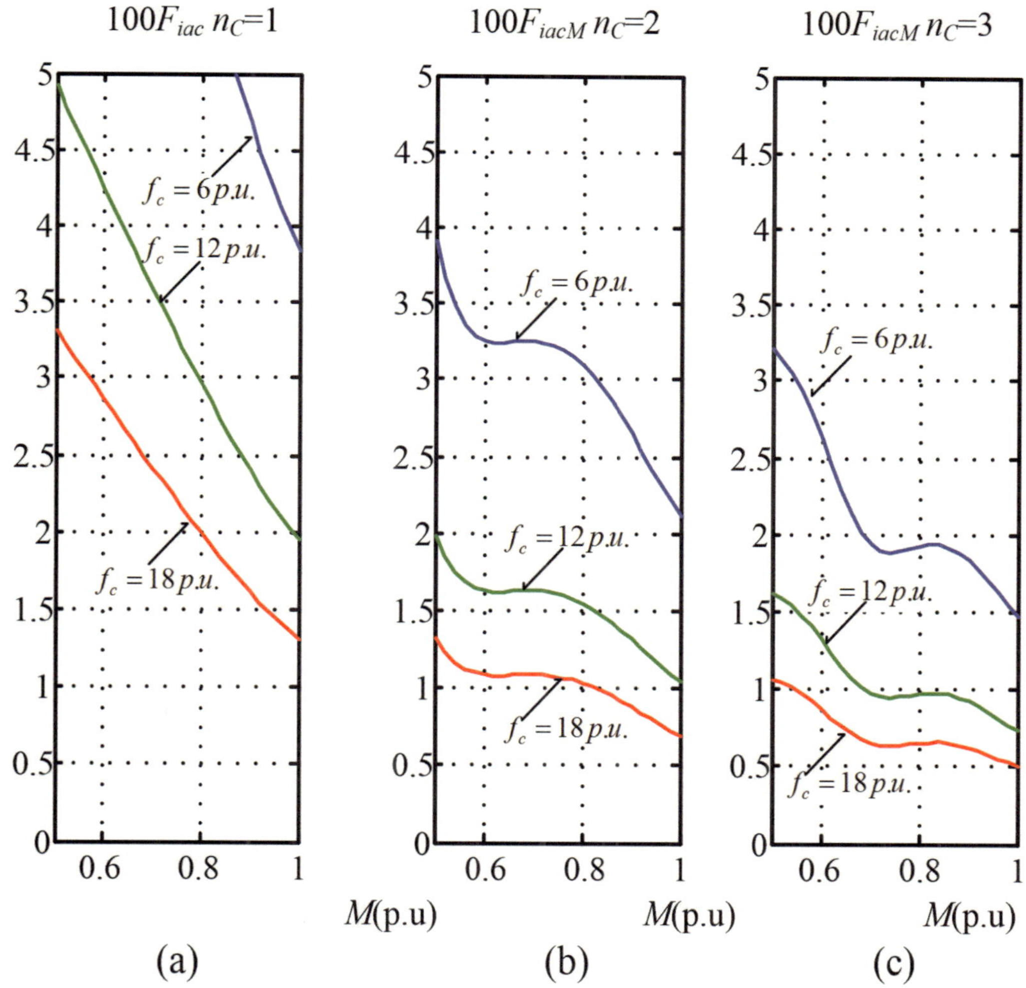

Figure 7. F_{iac} and F_{iacM} as functions of the inverters number and carrier frequency (a) F_{iac} for $n_C = 1$, 3 levels, (b) F_{iacM} for $n_C = 2$, 5 levels, and (c) F_{iacM} for $n_C = 3$, 7 levels.

used, then F_{iacM} is equal to n_C times F_{iac}. For example, a CHB-CSI with a unitary modulation index ($M_i = 1$), three cells per phase (n_C= 3), and a 6 p.u. carrier frequency ($F_{iacM} \approx 0.038$) shows $F_{iacM} \approx 0.114$.

5. AC drive application

5.1. Description

An example of the use of the CHB-CSI topology is AC Drive, where the power converter is connected in series to each phase of an electrical AC machine. In this case, a three-phase machine is fed by n_C cells by phase, so the AC drive has $3n_C$ cells as is shown in **Figure 8**. For this example, each cell is fed by a three-phase rectifier based on a current-source rectifier which is connected to the AC grid through a power transformer. This power transformer typically is a multistep transformer in classical CHB-CSI topology but can be simplified when active front ends are used, as in this case. Due to the use of a current-source rectifier, an LC filter is required at the cell input stage.

In the same way, the multicell topology is based on voltage-source inverters; the multicell topology based on current-source inverters is designed to increase the load voltage, which is the sum of the voltage on each cell. Due to the series connection, the current in each cell is the same. Then, the fundamental load current shown in **Figure 8** is given by

$$\vec{i}^{j}_{L,1} = \vec{i}_{oj1,1} = \vec{i}_{oj2,1} = \dots = \vec{i}_{ojn_C,1}. \tag{25}$$

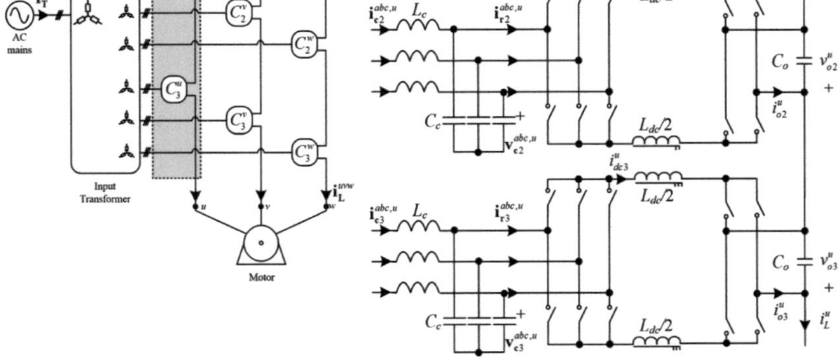

Figure 8. Cascaded H-bridge topology based on current-source inverter in an AC drive application.

The load voltage can be written in terms of the fundamental load current; thus,

$$\vec{v}^{\,j}_{L,1} = \vec{v}_{Nj,1} = \sum_{i=1}^{n_C} \vec{v}_{ji,1} = \vec{i}^{\,j}_{L,1}\vec{z}_{L,1}. \tag{26}$$

For the transformer input current, if there is not a phase shift between the primary and the secondary, one can write

$$\mathbf{i}^{abc}_{\mathbf{T}} = \frac{1}{n_T} \sum_{\substack{i=1 \\ j=u,v,w}}^{n_C} i^{abc}_{cji}, \tag{27}$$

and the cell input voltage is defined as

$$\mathbf{v}^{abc}_{\mathbf{sji}} = \frac{1}{n_T}\mathbf{v}^{abc}_{\mathbf{T}}. \tag{28}$$

Every cell is built up by a current source active front end which feeds a single-phase inverter through the DC link inductor. In order to obtain several controlled voltage sources connected in series array to the load, single-phase inverters and their respective capacitors are connected in series, achieving with this that each inverter-capacitor set behaves like a voltage source controlled through their DC currents. Then, each cell, as is shown in **Figure 9**, can be modeled in dq axis; thus

$$\mathbf{v}^{dq}_{\mathbf{r}} = L_C\left[\dot{\mathbf{i}}^{dq}_{\mathbf{r}} + \mathbf{W}\mathbf{i}^{dq}_{\mathbf{r}}\right] + \mathbf{v}^{dq}_{\mathbf{C}}, \tag{29}$$

$$\mathbf{i}^{dq}_{\mathbf{r}} = C_C\left[\dot{\mathbf{v}}^{dq}_{\mathbf{C}} + \mathbf{W}\mathbf{v}^{dq}_{\mathbf{C}}\right]\mathbf{v}^{dq}_{\mathbf{C}} + G_r\mathbf{m}^{dq}_{\mathbf{r}} i_{dc}, \tag{30}$$

$$\left[G_r\mathbf{m}^{dq}_{\mathbf{r}}\right]^T\mathbf{v}^{dq}_{\mathbf{C}} = L_{dc}\dot{i}_{dc} + s_i i_{dc}, \tag{31}$$

$$v_o = R_l i_o + L_s\frac{d}{dt}i_o, \tag{32}$$

where

$$\mathbf{W} = \begin{bmatrix} 0 & -2\pi f_r \\ 2\pi f_r & 0 \end{bmatrix} = \begin{bmatrix} 0 & -\omega_r \\ \omega_r & 0 \end{bmatrix}, \tag{33}$$

with f_r as the network frequency. Using the dq model, it is possible to define a control strategy for the rectifier stage. A control scheme must ensure the regulation of the DC current level in each cell (see **Figure 10**), allowing the use of a fixed modulation pattern in the inverter stage. The control scheme controls the active and reactive power using the currents at the input of the cell. An active power controller is used to control the DC current and, through the DC current, the cell output voltage, using the DC component of the output power of the cell. A reactive power controller can be used to compensate the reactive power of the LC filter. The control is

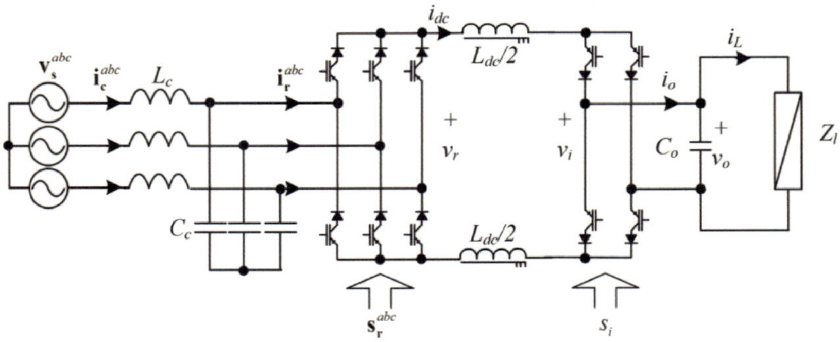

Figure 9. A power cell with an active front end stage.

replicated in each cell and the references of the output voltage, frequency, and cell reactive power are common to those controllers.

On the other hand, *dq* model allows to calculate the operation region of the topology and defines the active power P_{MC}, reactive power Q_{MC} and load voltage in terms of the number of cells n_C, the LC filter, the modulation at the rectifier stage, and the transformer voltage and its ratio; so

$$P_{MC} = V_T^d I_T^d = 3n_C \underbrace{\frac{V_T^{d^2}}{n_T} \frac{K_{dc} M_r^d}{\left(1 - \omega_r^2/\omega_{LC}^2\right)^2}}_{P_C} \approx P_{Load}, \tag{34}$$

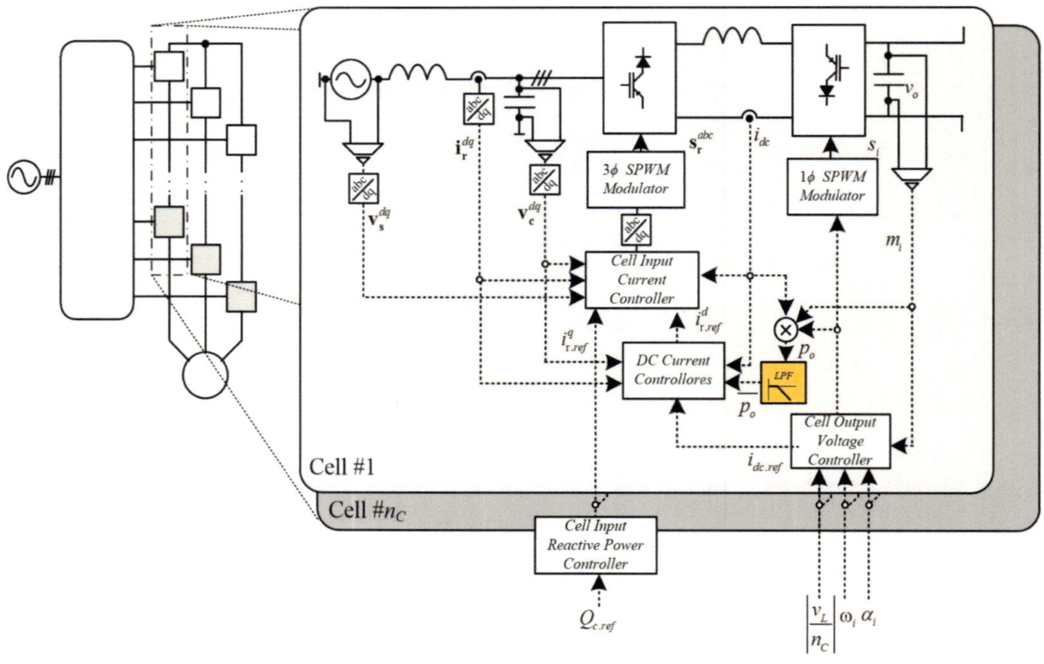

Figure 10. A control scheme for an AC drive based on CHB-CSI.

$$Q_{MC} = V_T^q I_T^d = \underbrace{3n_C \frac{V_T^{d^2}}{n_T} \left[\frac{\frac{G_r^2}{R_{dc}+R_i(\omega_i)} M_r^d M_r^q}{\left(1-\omega_r^2/\omega_{LC}^2\right)^2} + \frac{\omega_r C_c}{1-\omega_r^2/\omega_{LC}^2} \right]}_{Q_C}, \tag{35}$$

where P_c and Q_c are the active power and reactive power in each power cell, G_r is the gain of the modulation technique for the fundamental component, R_{dc} can be used to simulate the losses on the cell, and ω_{LC} is the resonance of the LC input filter given by

$$\omega_{LC} = \frac{1}{\sqrt{L_C C_C}}. \tag{36}$$

While the load voltage can be defined as

$$V_L(\omega_i) = n_C \frac{V_T^d}{n_T} \frac{M_r^d Z_m(\omega_i) \left(\frac{G_r^2}{R_{dc} + R_i(\omega_i)} \right)}{4 G_r \left(1 - \omega_r^2/\omega_{LC}^2\right)}, \tag{37}$$

and $R_i(\omega_i)$ is defined by

$$R_i(\omega_i) = \frac{1}{2n_C} Z_m(\omega_i) M_i^2 cos(\phi_m(\omega_i)). \tag{38}$$

from Eqs. (34) and (35), it is possible to notice that for a required active power on the load ($P_{MC} = P_{Load}$ only if $R_{dc} = 0$), it can be divided into $3n_C$ cells, and the same for the load voltage can be divided into n_C cells. On the other hand, from Eq. (36), it is possible to notice that the cell input voltage (i.e., the voltage at the secondary transformer) can be reduced when the cell number, n_C, is increased.

5.2. Examples

In order to show the performance of the CHB-CSI topology, an AC Drive is simulated using PSIM with the parameters shown in **Table 2**, considering a 9.33 MVA load per phase and a 0.8 inductive power factor, using one cell per phase (n_C= 1) and two cells per phase (n_C= 2).

Figure 11 shows the operating region for n_C= 1 and n_C= 2 as functions of the DC current level, including active and reactive power per cell (**Figure 11a**) and load voltage and inverter voltage per cell (**Figure 11b**), where the RMS cell input voltage for $n_C = 1$ is equal to 4 kV and, in order to get the same load voltage level, for n_C= 2, the RMS cell input voltage is reduced to 2 kV—reducing the semiconductor voltage rating with respect to $n_C = 1$).

In terms of steady-state performance, **Figure 12** shows the key waveform $n_C = 1$ and $n_C = 2$ using a DC current level equal to 500 A per cell (**Figure 12a**) and a load frequency equal to 50

Parameter		Value		Value p.u.
Load	R_l	20	Ω	0.804
	L_l	47	mH	0.593
Voltage at the transformer primary	V_{Trms}	8	kV	
Transformer ratio for $n_C = 1$	nT	2		
Transformer ratio for $n_C = 2$	nT	4		
Lc Filter at the cell input	L_c	12	mH	0.15
	C_c	40	μF	3.2
DC inductor	L_{dc}	15·36	mH	6.82
Output capacitor	C_o	20	μF	6.4
Switching frequency for rectifier stage		750	Hz	15
Switching frequency for inverter stage		500	Hz	10

Table 2. AC drive parameters.

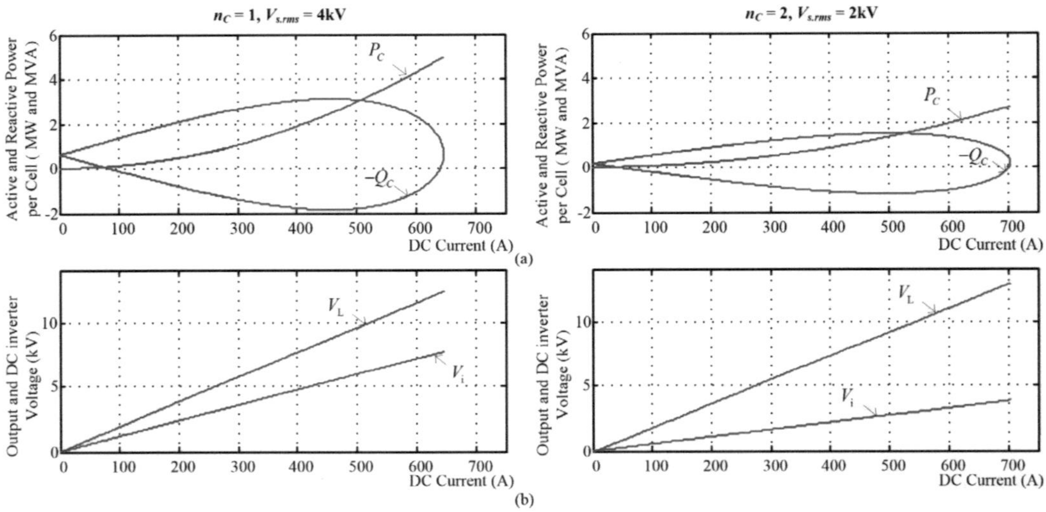

Figure 11. Operation region for $n_C = 1$ and $n_C = 2$ (a) input active and reactive power, per cell and (b) load voltage and inverter output voltage.

Hz. For this case, unitary displacement power factor at the input of the power converter has been set in order to get input current in phase with the input voltage. It is possible to notice that, for the same DC inductor parameters, the ripple by the oscillating power is lesser in $n_C = 2$ than in $n_C = 1$—near to 50% less, while in both cases, the average value of the DC current is 500 A. About the load voltage, **Figure 12b** shows that both cases reach similar voltage levels, but the harmonic distortion in $n_C = 2$ is lesser than $n_C = 1$ because of the use of a multilevel modulation technique. About the quality in the input of the cell and the input of the power

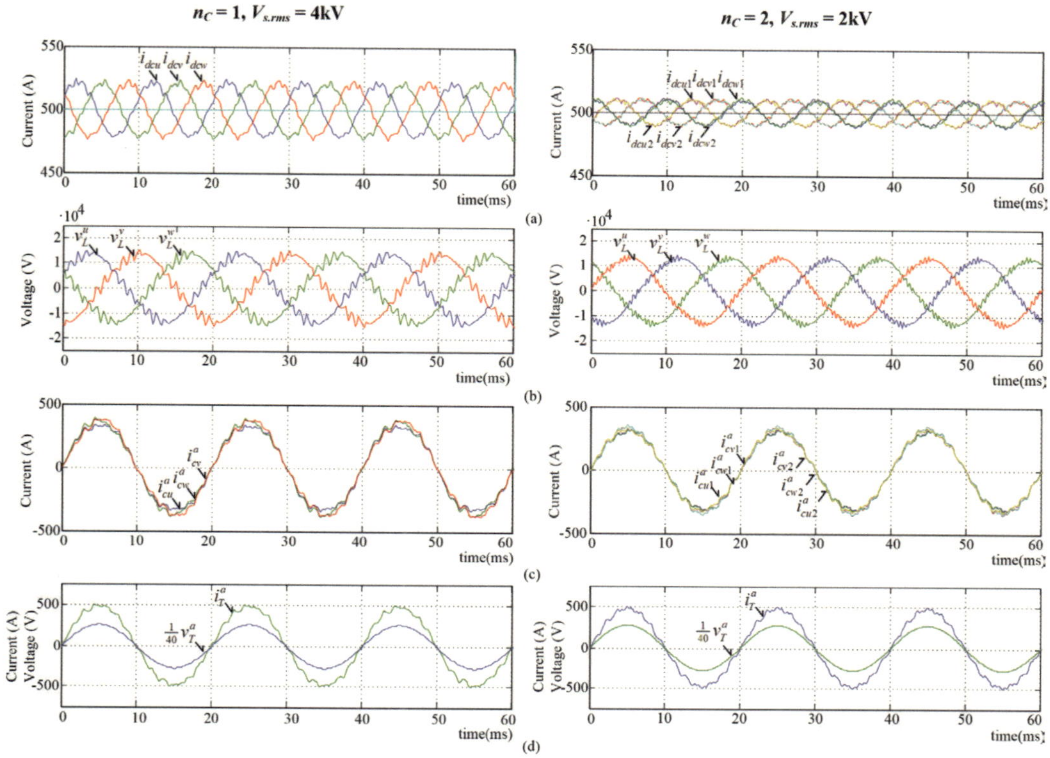

Figure 12. Steady state key waveform for n_C= 1 and n_C= 2 (a) DC current in the cells, (b)load voltage, (c)cell input currents, and (d) transformer input current and transformer primary voltage.

transformer, **Figure 12c** and **d** shows voltage and current for both cases, where one can see the low distortion in the input current—typically in current-source rectifier topologies.

Figure 13 shows the frequency spectra for the DC current (**Figure 13a**), load voltage (**Figure 13b**), and cell input current (**Figure 13c**). From the DC current frequency spectra, a second harmonic can be noticed due to the oscillating power drained by the single-phase inverter limited by the DC inductor design. This component is lesser in the n_C =2 than n_C = 1, but the effect in the load voltage—third harmonic—is similar. Finally, at the cell input current, there is a third harmonic caused by the second harmonic in the DC current. This harmonic does not exist at the transformer input current because it is compensated among cells that feed different phases of the load.

Finally, for n_C = 1 and n_C =2, the performance under load frequency changes and DC current level change has been tested (**Figures 14** and **15**, respectively). For these tests, a non-linear control has been implemented in order to control the load voltage using the DC current level, which is controlled by the rectifier stage. In the first case, under frequency changes from 20 to 70 Hz (**Figure 14b**), the power topology is able to impose it on the load. For lower frequencies, the amplitude of the second harmonic in the DC current increases because the DC inductor has been designed to limit it at 50 Hz (**Figure 14a**) and decreases for higher frequencies. On the other hand, at the input current (**Figure 14c**), it is possible to notice the effect of the second harmonic in the DC current. This oscillation is not presented in the current at the primary transformer because of the compensation of these components among cells that feed different load phases.

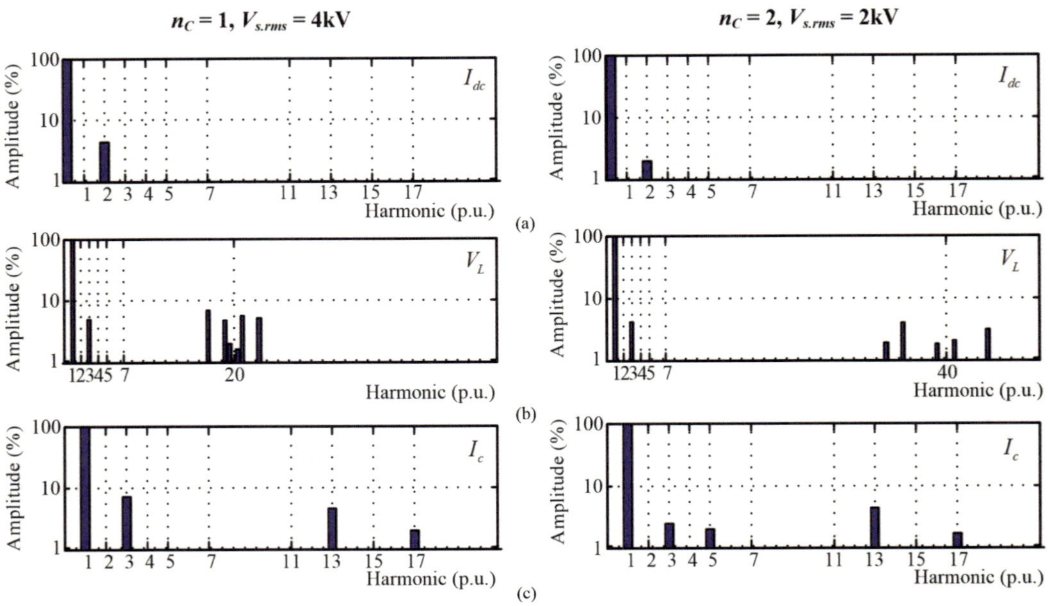

Figure 13. Frequency spectra for $n_C = 1$ and $n_C = 2$ (a) DC current, (b) load voltage, and (c) cell input current.

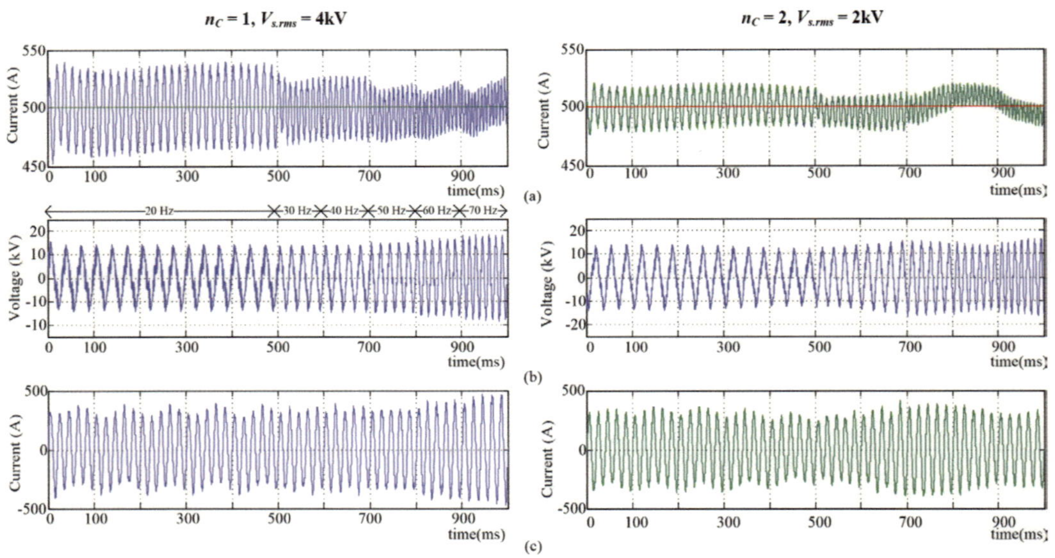

Figure 14. Response for load voltage frequency changes (a) DC current, (b) load voltage, and (c) cell input current.

About the DC current step change, **Figure 15a** shows a 10% step in t = 100 ms. In both cases, it can be notice that the load voltage increases by 10% (**Figure 15b**), while the input current increases in the same rate due the increases in the load power (**Figure 15c**). In this case, the dynamic is defined by the controller parameters and can be specified in the controller design process.

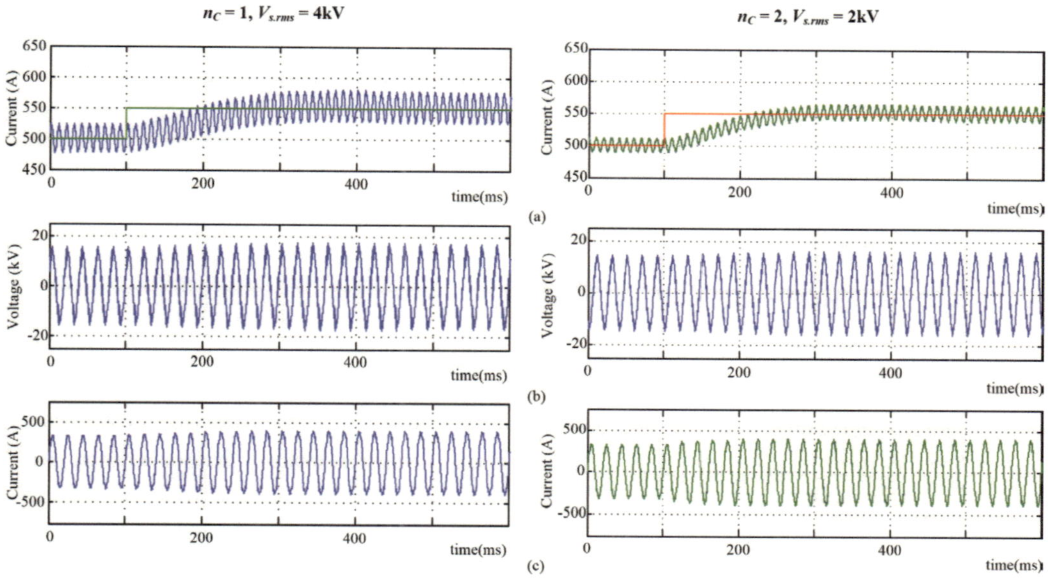

Figure 15. Key waveform for 10% DC current step (a) DC current, (b) load voltage, and (c) cell input current.

6. Conclusions

Cascaded H-bridge topologies based on CHB-CSI are emerging topologies that use the same principle of a cascaded H-bridge converter, allowing to divide the required load voltage level and power into several single-phase inverters connected in series. Advantages of the proposed topology are (i) high quality of voltage and current waveforms using lower switching frequencies and (ii) inherent short-circuit protection because of the use of current-source inverter, while its main drawbacks are (i) the use of a bulky DC inductor because of the use of current-source inverters and (ii) the oscillating power drained by the inverter on the DC side, because the use of single-phase inverters. With an appropriate control scheme, the CHB-CSI is able to impose a desire frequency and load voltage level. In case of AC drive applications, an increase in the number of cells allows reducing the voltage rating of the components without reducing the operation region of the whole converter. At the same time, the DC current variation in each cell decreases when the number of cells increases. On the other hand, load voltage can be regulated through the DC current control, allowing the use of a fixed modulation index for the inverter stage. The above allows designing the capacitive filter with the minimum F_{iacM} required for a given modulation technique and switching frequency.

Acknowledgements

This work was funded by Fondecyt Chile under Grant 111 40759. The financial support of the research group *Electrical Power Conditioning and Conversion*—GI160510 EF—of the Universidad del Bío-Bío and the technical support of the Power Conditioning and Conversion Laboratory of the same university are also acknowledged.

Nomenclature

n_C	Number of cells in a series array
Z_L	Load impedance
s_i	Modulation function of the inverter
$\mathrm{v_o}$	Cell output voltage
$\mathbf{v_l}$	Load voltage
$\mathrm{i_o}$	Cell output current
$\mathbf{i_L}$	Load current
i_{dc}	DC current
s_o	Cell apparent output power
$\mathbf{v_s}$	Cell voltage supply
$\mathbf{i_c}$	Cell input current
$\mathbf{s_r}$	Modulating vector of the rectifier
$\mathbf{i_c}$	Cell input current
abc	Stationary coordinates for three-phase input
uvw	Stationary coordinates for three-phase output
dq	Rotating coordinates

Author details

Pedro Eduardo Melín Coloma[1]*, José Rubén Espinoza Castro[2], Carlos Rodrigo Baier Fuentes[3] and Jaime Addin Rohten Carrasco[1]

*Address all correspondence to: pemelin@ubiobio.cl

1 Department of Electrical and Electronic Engineering, Bio-Bio University, Concepción, Chile

2 Department of Electrical Engineering, Concepcion University, Concepción, Chile

3 Department of Electromechanics and Energy Conversion, Talca University, Curicó, Chile

References

[1] Bin W, editor. High-power converters and AC Drives. 1st ed. USA: Wiley, IEEE Press; 2017. p. 480. DOI: 10.1002/9781119156079

[2] Bin W, Pontt J, Rodriguez J, Bernet S, Kouro S. Current-source converter and cycloconverter topologies for industrial medium-voltage drives. IEEE Transactions on Industrial Electronics. 2008;**55**(77)

[3] Rodriguez J, Bernet S, Bin Wu, Pontt JO,; Kouro S. Multilevel voltage-source-converter topologies for industrial medium-voltage drives. IEEE Transactions on Industrial Electronics. 2007;**54**(6)

[4] Franquelo LG, Rodriguez J, Leon JI, Kouro S, Portillo R, Prats MAM. The age of multilevel converters arrives. Industrial Electronics Magazine, IEEE. 2008;**54**(6)

[5] Teodorescu R, Blaabjerg F, Pedersen JK, Cengelci E, Enjeti PN. Multilevel inverter by cascading industrial VSI. IEEE Transactions on Industrial Electronics. 2002;**46**(4)

[6] Sanchez-Ruiz A, Mazuela M, Alvarez S, Abad G, Baraia I. Medium voltage–high power converter topologies comparison procedure, for a 6.6 kV drive application using 4.5 kV IGBT modules. IEEE Transactions on Industrial Electronics. 2012;**59**(3)

[7] Malinowski M, Gopakumar K, Rodriguez J, Pérez MA. A survey on cascaded multilevel inverters. IEEE Transactions on Industrial Electronics. 2010;**57**(7)

[8] Ko Y, Andresen M, Buticchi G, Liserre M. Power routing for cascaded H-bridge converter. IEEE Transactions on Power Electronics. Forthcoming. DOI: 10.1109/TPEL.2017.2658182

[9] Melin PE, Espinoza JR, Zargari NR, Moran LA, Guzman JI. A novel multi-level converter based on current source power cell. In: IEEE, editor. Power Electronics Specialists Conference, 2008

[10] Melin PE, Espinoza JR, Baier CR, Rohten JA, Ramirez RO, Moran LA. Improved control scheme towards reduced DC link inductors in a multi-cell topology based on current source converters. In IECON 2012, 38th Annual Conference on IEEE Industrial Electronics Society; October 25–28, 2012; Montreal IEEE, USA

[11] Baier CR, Espinoza JR, Melin PE, Espinoza E, Munoz J. A novel multi-level CSI based topology with inter-cell magnetic couplings for minimum DC storage components. In: Industrial Electronics (ISIE), 2010 IEEE International Symposium; Bari, Italia. July 4–7, 2010

[12] Melin PE, Baier CR, Espinoza JR, Munoz JA, Ramirez RR, Maurelia EA. On the DC inductors size reduction in a multi-cell topology based on current source converters by means of magnetic couplings. In: Industrial Electronics Society, IECON 2013, 39th Annual Conference of the IEEE; Wien, Austria. November 10–13, 2013; IEEE, USA

[13] Baier CR, Melin PE, Guzman JI, Rivera M, Muñoz JA, Rohten J, Espinoza J. Current-source cascaded multilevel converters based on single-phase power cells. In: Industrial

Electronics Society, IECON 2013, 39th Annual Conference of the IEEE; Wien, Austria. November 10–13, 2013; IEEE, USA

[14] Melin PE, Espinoza JR, Rohten JA, Espinosa EE, Baier CR, Guzman JI. Cascaded H-bridge topologies comparison for multi-cell current-source inverters under different DC inductor size reduction methods. In: Industrial Electronics Society, IECON 2014, 40th Annual Conference of the IEEE; October 29, 2014–November 1, 2014; Dallas. IEEE, USA

[15] Rohten J, Melin P, Espinoza J, Silva J, Espinosa E, Munoz J, Sbarbaro D. Resonant control for H-Bridge topologies based on single-phase current-source inverters. In: Industrial Electronics (ISIE), 2014. IEEE 23rd International Symposium; Istanbul, Turkey, June 1–4, 2014. IEEE, USA

[16] Melin P, Espinoza J, Silva J, Espinosa E, Munoz J, Baier C. Resonant control for multi-cell cascaded H-bridge topologies based on current source inverters. In: Industrial Electronics Society, IECON 2014, 40th Annual Conference of the IEEE; October 29, 2014–November 1, 2014; Dallas. IEEE

[17] Carlos R Baier, Miguel A Torres, Javier Muñoz, Johan I. Guzman, Pedro E. Melin, Jaime Rothen, Marco Rivera. d-q-DC reference frame control strategy for single-phase current source cascaded inverters. In: IECON 2014, 40th Annual Conference of the IEEE Industrial Electronics Society; Dallas. USA: IEEE;2014. DOI: 10.1109/IECON.2014.7049198

[18] Melin P, Espinoza J, Guzman J, Rivera M, Espinosa E, Rohten J. Analysis and design of a cascaded H-bridge topology based on current-source inverters. In: Industrial Electronics Society, IECON 2013, 39th Annual Conference of the IEEE. Wien, November 10–13, 2013

[19] Pedro E Melin, Jaime A Rohten, Jose R Espinoza, Carlos R Baier, Eduardo E Espinosa, Javier A Muñoz, Javier A Riedemann. Analysis and design of a multicell topology based on three-phase/single-phase current-source cells. IEEE Transactions on Power Electronics. 2016;**31**(9):ISI. DOI: 10.1109/TPEL.2015.2502185

Chapter 4

Pulse Density Modulation Applied to Series Resonant Inverter and Ac-Ac Conversion

Abdelhalim Sandali and Ahmed Chériti

Additional information is available at the end of the chapter

http://dx.doi.org/10.5772/intechopen.68324

Abstract

The PDM control joins together between the concepts of soft switching and hard switching. Its application to the series resonant inverter cancels the switching losses and uses dc bus without storage capacity. Objectively, the PDM controls led to ac-ac converters with high efficiency (zero switching loss), small size (no storage capacity) and with the possibility of a self power factor correction. However, the operating analysis of these converters is very complex because the operation is done on two time scales and leaves questions unanswered. The average modeling facilitates the analysis of the operation and leads to establish: (i) an analytical expression of the power factor, (ii) the linearity conditions of the power characteristic, and (iii) a model of ac-ac series resonant multi-converter which is independent of the carriers. In the case of ac-ac series resonant multi-converter, the coordination of carriers allows to shape the power characteristic. Among the three types of coordination presented, there is an original coordinate that linearizes the power characteristic. The results are validated by simulations carried out in Matlab SimPower systems.

Keywords: pulse density modulation, series resonant inverter, multi-inverter ac-ac converter, power factor correction, average modeling

1. Introduction

The soft-switching, appeared in the early eighties of last century, is a major event in the development of power electronics [1, 2]. Numerous research and conference sessions devoted to it reflect this importance. The soft switching is a conceptual breakthrough that led to technological advances. Indeed, to reduce the size of the reactive components, something that has a positive effect on the weight and size of converters, it is necessary to seek to increase the switching frequency. However, to increase the switching frequency of power semiconductor switches, it is imperative to create the conditions for reducing switching losses. Within the framework of soft switching concept, one considers that the most effective way to achieve this goal is to leave full-controlled power semiconductor devices to switch depending on their changing voltages and/or currents [3–5]. The power part of the converter plays an active role in determining the switching instants.

In resonant converters, implementing this concept, there is a shared determination of switching instants between the control part and power part of the converter: if a commutation is caused by the control, complementary commutation is caused by the voltage or current of the switch (ZVS or ZCS). The transmitted power control is done by the frequency modulation.

Pulse density modulation (PDM) control, appeared in the mid-1990s [6, 7], joins together the concept of soft switching and the traditional concept of hard switching by separating the roles of the control part and power part. The power part is responsible for determining the switching instants. The control part decides the nature of switching cycles (active or inactive). Its application to series resonance inverters has the major advantage to cancel the switching losses and to produce an output power factor near to unity [6–11]. The integration of these inverters in the ac-ac conversion makes it possible to save the smoothing filter and to have a sine-wave absorption at full power [6, 7]. Several recent researches are devoted to the development of PDM control and to the valorization of its applications [8–11]. The operating analysis presented in these papers focus on the output current. The input current (current drawn from the ac-supply) analysis is forsaken. This aspect constitutes the poor relation in the scientific literature dedicated to the PDM technique.

Fill this blank, clarify it why, and show how to exploit the benefits and manage the challenges are the objectives of this work. More than a synthesis of previous work, this chapter provides for the first time an average modeling of PDM inverters, an accurate determination of the linear operating conditions and an original linearization technique.

This chapter is organized as follows: in Section3, we present the principle of PDM control and its integration in ac-ac converter. Section4 is devoted to the description of a pulse density modulator. The conventional analysis is the subject of Section5. The determinations of the input current of the ac-ac converter and its power factor in the case of single and multi-inverter configurations are presented. Section6 is devoted to the average modeling of ac-ac series resonant converter in single- and multi-inverter configurations. Several cases of coordination of the carriers are discussed in Section7. Simulation results are given in Section8, and conclusion is presented in Section9.

2. Principle

Pulse density modulation (PDM) is a type of control applied to the series resonance inverters. Turn-on and turn-off occur at zero crossings of the load current, because the switching frequency is taken equal to resonance frequency of the load. All commutations are lossless and without current gradient, and the input current of inverter is unidirectional. The inverter operation has the following improvements: (i) the switches are completely released from switching

http://dx.doi.org/10.5772/intechopen.68324

stress; (ii) dc bus has no storage capacity; and (iii) reduction of electromagnetic noise. But, as the switching frequency is now fixed, it is impossible to use it to control the power.

To avoid this disadvantage, the control part generates a PDM pattern that determines whether a switching cycle is active or inactive. An active cycle is a normal operating cycle of the single-phase inverters. An inactive cycle is defined by the simultaneous state-on of same side switches (e.g., high side) and the simultaneous state-off of the other side switches (e.g., low side). It puts the output inverter in freewheel and, consequently, interrupts the flow of energy between input and output of the converter. The power control is now done by the PDM pattern duty cycle defined by:

$$d = n/k \tag{1}$$

with k is the number of total cycles per PDM pattern period (called PDM pattern length) and n=1, 2, ... or k.

The power control is thus done in a discrete manner with a resolution which depends on the length of the pattern.

Figure 1 shows the considered inverters and the transcription logic circuit of the PDM pattern in the gate control signals of the switches. We propose in **Figure 2** an operating model that clearly shows the coexistence of the two concepts (hard and soft switching). The PDM inverter is divided into soft inverter and hard buck. The transmitted power is controlled by the PDM pattern.

When the inverter is supplied by a single-phase diode bridge, the unit forms an ac-ac converter (LF to HF) (**Figure 3**). Since the input current of PDM inverter is unidirectional, it is possible to eliminate the smoothing filter (low frequency 2×50 or 2×60) and keep only a high frequency decoupling capacitor C_{HF}. The latter absorbs the high frequency ripple of the current in the

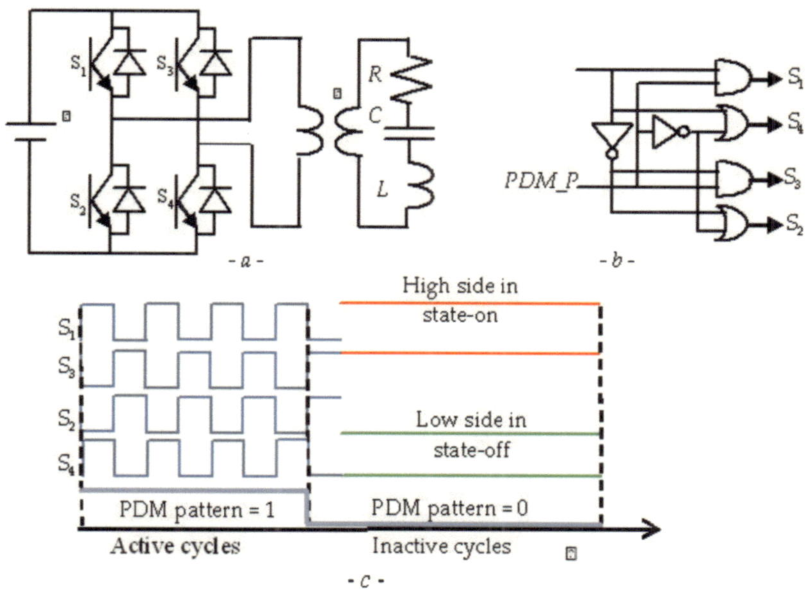

Figure 1. PDM inverter: topology, control and definition.

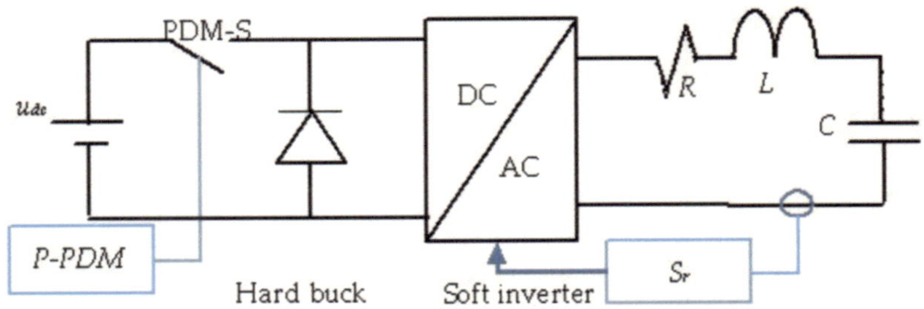

Figure 2. Decomposition of the PDM inverter in hard buck and soft inverter.

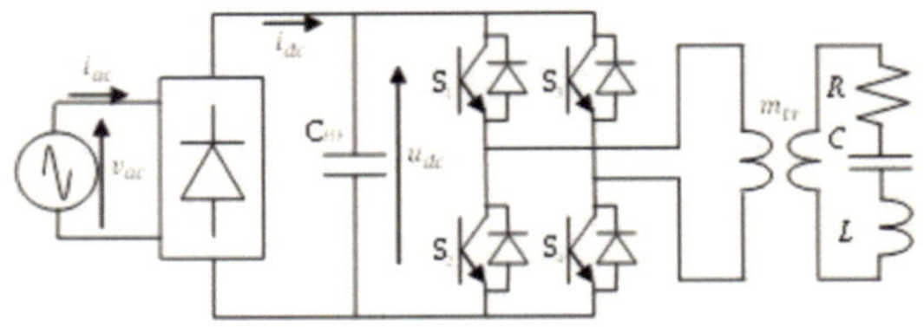

Figure 3. Ac-ac PDM converter (without smoothing filter).

rectifier. The result is a small-sized ac-ac converter because it is relieved of the smoothing filter). The current drawn by the ac-ac converter is a sinusoidal form, but is intersected by zero current phases when inactive cycles occur. In the following, we focus on the control, the transmitted power and the quality of current drawn by this ac-ac converter that will be called ac-ac PDM converter and noted PDMC.

3. PDM pattern generation

The PDM pattern generation is subject to two particular restrictions.

The first restriction concerns the adaptation of the PDM pattern frequency to the load resonance frequency. So that the switching cycles are not truncated, the PDM pattern period must be a multiple of the resonance period:

$$\text{First restriction } T_{PDM} = kT_r \text{ with } k \text{ is an integer} \tag{2}$$

To satisfy this restriction, several options are possible. We present below a PWM type technique but adapted to PDM control. It is based on a synchronous comparison of a control signal (e_c) and a triangular carrier (Car) [12]. The synchronous comparison is carried out by a conventional comparator followed by a D flip-flop (**Figure 4**).

When the PDM inverter is integrated into an ac-ac series resonant converter, it is necessary that the PDM pattern satisfies a second restriction: its period (frequency) must be a sub-multiple (multiple) of the period (frequency) of the dc link voltage:

http://dx.doi.org/10.5772/intechopen.68324

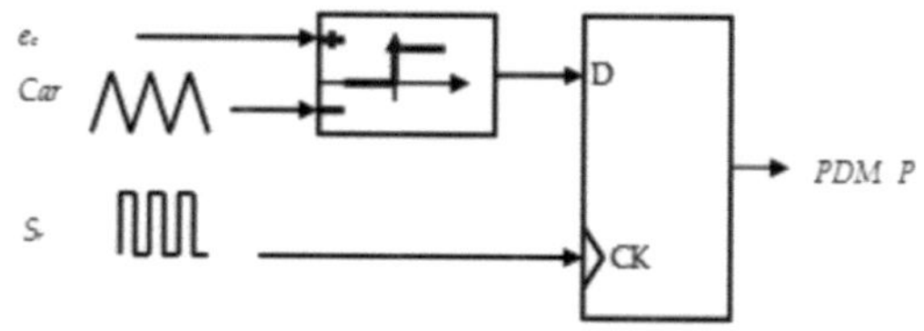

Figure 4. Adaptation of PDM frequency to resonance frequency by synchronous comparison.

$$\text{Second restriction } T_{dc} = qT_{PDM} \text{ with } q \text{ is an integer} \tag{3}$$

If this restriction is not satisfied, a continuous component can appear in the current drawn from the electrical network. The substitution of Eq. (2) into Eq. (3) leads to restriction:

$$T_{dc} = NT_{res} \text{ with } N = q * k \text{ is an integer} \tag{4}$$

To check this restriction, a calibration of the dc-voltage period seen by the inverter is performed by the definition of the ZCD signal (Zero Crossing Detector) [12]. This signal is determined by the synchronous comparison of the ac-supply voltage with positive and negative thresholds. The synchronous comparison is carried out here by the set two comparators—and logical gate—D flip-flop. ZCD is at the zero level during a time which is a multiple of the period of resonance and which can be very close to the period of the rectified voltage if the thresholds are close to zero. ZCD, when it's high, is used to reset the PDM pattern and the carrier's generator. Henceforth, the useful (or usable) period of the rectified voltage is automatically adjusted with a multiple of the resonance period, and the triangular carrier is synchronized to the latter. **Figure 5** shows the pattern generation circuit for an ac-ac PDM converter. The PDM pattern duty cycle is identified with the control signal because the carrier is triangular, whereas the carrier frequency represents the PDM pattern length. As a result, the control signal controls the power while the carrier frequency controls the power variation resolution. In the following section, we establish the relation between the control signal and the transmitted power and the impact of the carrier frequency on the quality of the current drawn from the ac-supply.

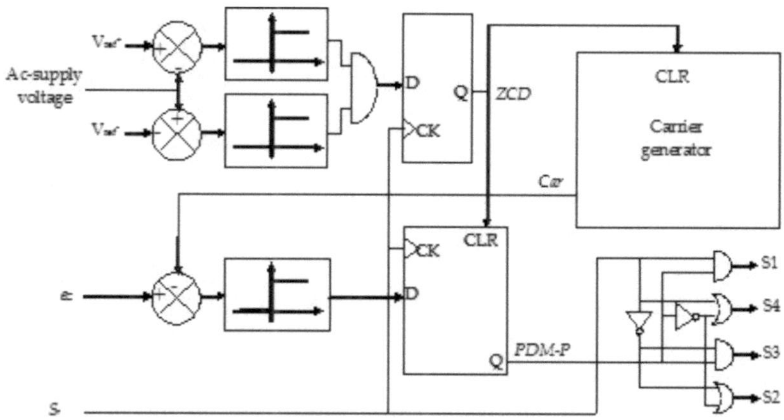

Figure 5. PDM pattern generator performing synchronous comparison and calibration.

4. Conventional analysis

4.1. Input current of ac-ac PDM converter

The drawn current from ac network and transmitted power by ac-ac series resonant converter are determined under the following assumptions: (i) the load has a low damping coefficient; (ii) C_{HF} capacitor absorbs HF component (twice the resonant frequency) of the inverter input current; (iii) the dc link voltage is assumed constant during a switching cycle; and (iv) no restriction is imposed on the PDM pattern except the restrictions of definition.

One determines successively the load current, the currents after and before the HF decoupling capacitor and the current drawn from the ac-supply. Then, one calculates the Fourier series of the latter current. In Ref. [12], it is shown that if the PDM pattern is generated according to the solution of Section3, the following results are obtained:

i. the transmitted power in pu varies linearly with the control signal:

$$p = e_c \tag{5}$$

ii. the spectrum of the drawn current consists of harmonic pairs spaced from $2qF_{ac}$

iii. and the power factor, calculated from the harmonic summation, is as follows:

$$PF = 1 \Bigg/ \sqrt{1 + \frac{2}{{e_c}^2} \sum_{h=1}^{\infty} \Big(sin\big(\pi h(1 - e_c)\big) / \pi h \Big)^2} \tag{6}$$

It is independent of the PDM pattern frequency. The power factor decreases continuously when the transmitted power in pu varies from 1 to 0.

Failing to improve the power-factor directly, an increase in the pattern frequency brings a better conditioning of the harmonic distortion (increase spacing and thus reject harmonics in high frequencies). But, it is observed that the more the frequency increases, the more one loses the linearity between power and control signal. We propose later in this chapter, a theoretical determination of the maximum frequency which preserves this linearity.

4.2. Input current of ac-ac multi-PDM converter and power factor correction

Correction by mutual compensation requires the use more than one inverter and an adapted control. The inverters are managed in such a way that the distortion produced by an inverter (zero current phase) is completely or partially masked by the other inverters. The inverters do not operate in inactive cycles simultaneously but successively. This management is based on the use of a set of interlaced carriers. This idea was developed for a system with several inverters and separate loads (each inverter feeds one load). Then, the idea was extended to the more realistic case of a single load [13]. The considered converter and its carriers are shown in **Figure 6**. It is called ac-ac multi-PDM converter and noted M-PDMC$_G$. If this converter consists of G inverters (Inv$_g$ with $g = 1, 2, \ldots, G$), the carrier associated with an Inv$_g$ is as follows:

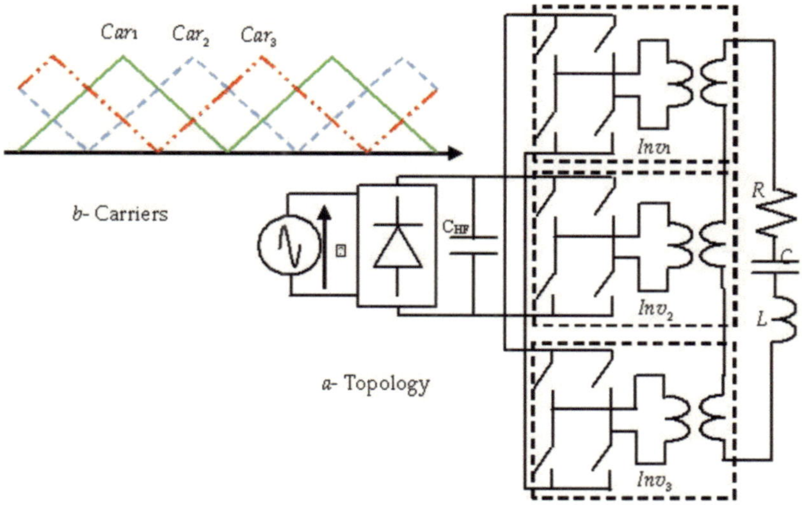

Figure 6. Ac-ac multi-PDM converter ($MPDMC_G$): topology and carriers in case $G=3$.

$$Car_g(\theta) = Car_1\left(\theta - (g-1)2\pi/G\right) \tag{7}$$

and all inverters have the same control signal:

$$e_{c,g} = e_c \tag{8}$$

This converter behavior is modeled by a bi-converter system with separate loads ($Conv_A$ and $Conv_B$) and variable parameters. The control signals, carriers and transformer ratios of $Conv_A$ and $Conv_B$ vary according to the control signal. **Figure 7** shows the topology of the bi-converter system and the control parameters. The results, detailed in Ref. [13], are as follows:

i. Power versus control signal is piecewise linear:

$$p = a_1 = g + (G.e_c - g)(2g-1) \tag{9}$$

Where $g - 1 = floor(G.e_c)$ (integer portion of $G.e_c$)

ii. the spectrum of the drawn current consists of harmonic pairs spaced from $2GqF_{ac}$

iii. the power factor, calculated from the harmonic summation, is:

$$PF = 1\Bigg/\sqrt{1 + 2(2g-1)^2 \frac{\sum_{h=1}^{\infty}\left(sin(\pi hGe_c)/\pi h\right)^2}{\left(g^2 + (Ge_c - g)(2g-1)\right)^2}} \tag{10}$$

The power factor is equal to 1 in G points when the transmitted power is equal to $(j/G)^2$ 100% of its maximum value with $j = 1,\ 2,\ \ldots G$, and

iv. the maximum power is G^2 times greater than in the case of the PDMC.

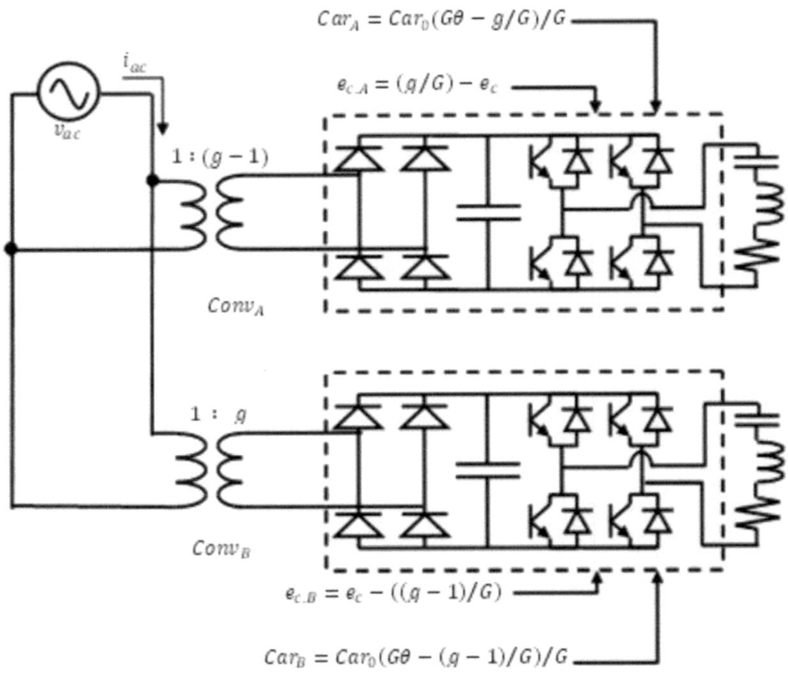

Figure 7. Bi-converter equivalent system when $(g-1)/G \leq e_c \leq g/G$.

This modeling is empirical because it is based solely on the observation of the converter behavior. However, since the behavior of a converter is defined by its topology and its control, the bi-converter model is only valid for the control law considered during the observation phase.

5. Analysis by average modeling

The average modeling is a powerful tool for the regulation and simulation of power electronics converters. Its main disadvantage is that it can predict only the mean value [14]. High frequency ripple is lost or predicted with inaccuracy [15]. In the case of PDM converters, the dc link between the rectifier and the inverter ensures high frequency decoupling. The rectifier's currents do not have high frequency ripple. The average modeling is thus a well-adapted tool for the determination of the low frequency side currents in the PDM converters. This technique is already applied successfully to the PDM-dual converters [16]. In the present work, it is applied for the first time to the PDM converters.

5.1. Average model of ac-ac PDM converter

5.1.1. Modeling of series resonant inverter

The time is subdivided into half-periods of resonance. Each half-period is indicated by an index n. The load voltage switches from $-u_{dc}$ to $+u_{dc}$ at instants $n(T_r/2)$ and from $+u_{dc}$ to $-u_{dc}$ at instants $(n+1)(T_r/2)$ with n even number. The capacitor voltage is as follows:

http://dx.doi.org/10.5772/intechopen.68324

$$v_C(n+1) = -v_C(n)\,A + u_{dc} * (1+A) \text{ at the end of the } (n+1)\text{th half period} \tag{11}$$

$$v_C(n+2) = -v_C(n+1)\,A - u_{dc} * (1+A) \text{ at the end of the } (n+2)\text{th half period} \tag{12}$$

Under compact form, Eq. (11) and Eq. (12) become:

$$v_C(n) = -v_C(n-1)\,A + (-1)^{n+1} u_{dc} * (1+A) \text{ with } n \text{ even or odd} \tag{13}$$

Eq. (13) gives the value of the capacitor voltage at the end of the nth half-period as a function of the value taken at the $(n-1)$th half-period. According to the initial value, Eq. (13) becomes:

$$v_C(n) = (-1)^n \left\{ v_C(0)\,A^n - u_{dc} * (1+A) \sum_{i=0}^{n-1} A^i \right\} \tag{14}$$

Knowing that $\sum_{i=0}^{n-1} A^i = \frac{1-A^n}{1-A}$, Eq. (14) becomes :

$$v_C(n) = (-1)^n \left\{ v_C(0)\,A^n - u_{dc} * (1-A^n)\frac{1+A}{1-A} \right\} \tag{15}$$

The pic value of load current during the nth half-period is as follows:

$$\hat{i}(n) = \sqrt{C/L}\left(v_C(n) - u_{dc}\right) exp(\alpha T_r/4) \text{ for } n \text{ odd} \tag{16}$$

$$\hat{i}(n) = -\sqrt{C/L}\left(v_C(n) + u_{dc}\right) exp(\alpha T_r/4) \text{ for } n \text{ even} \tag{17}$$

During the nth half-period, the average value of inverter's input current is as follows:

$$\langle i \rangle(n) = \frac{2}{\pi}\hat{i}(n) = \frac{2}{\pi}\sqrt{C/L}\left((-1)^{n+1} v_C(n) - u_{dc}\right)\cdot exp(\alpha T_r/4) \tag{18}$$

Since the initial value of u_{dc} is zero, writing Eq. (18) for $n=0$ leads to:

$$\langle i \rangle(0) = -\frac{2}{\pi}\sqrt{C/L} v_C(0)\cdot exp(\alpha T_r/4) \tag{19}$$

It is deduced that

$$v_C(0) = -\frac{\pi}{2}\sqrt{C/L}\langle i \rangle(0)\cdot exp(-\alpha T_r/4) \tag{20}$$

The substitution of Eq. (20) into Eq. (15) yields:

$$v_C(n) = (-1)^n \left\{ -\frac{\pi}{2}\sqrt{C/L} exp(-\alpha T_r/4)\langle i \rangle(0)\,A^n - u_{dc} * (1-A^n)\frac{1+A}{1-A} \right\} \tag{21}$$

The substitution of Eq. (21) into Eq. (18) yields:

$$\langle i \rangle(n) = \langle i \rangle(0)\, A^n + u_{dc} \frac{4}{\pi} \sqrt{C/L} \frac{exp(-\alpha T_r/4)}{1-A} \left\{ 1 - A^n \frac{1+A}{2A} \right\} \tag{22}$$

Under the assumption:

$$(1+A)/2A \approx 1 \tag{23}$$

(this assumption is justified by the fact that $(\alpha T_r/2 = \pi\xi)$ is close to zero), expression (22) is identified with that of the current in $R_{eq} - L_{eq}$ series branch supplied by a voltage u_{dc} with:

$$R_{eq} = \frac{\pi}{4} \sqrt{\frac{L}{C}} \frac{1 - exp(-\alpha T_r/2)}{exp(-\alpha T_r/4)} \tag{24}$$

$$L_{eq} = \frac{R_{eq}}{\alpha} = L \frac{\pi}{4\xi} \frac{1 - exp(-\alpha T_r/2)}{exp(-\alpha T_r/4)} \tag{25}$$

The inverter dc side can then be modeled by the average circuit constituted by $R_{eq} - L_{eq}$ series branch.

5.1.2. Modeling of PDM inverter

Pulse density modulation is introduced by considering a fictitious hard buck that connects the voltage source to the inverter, represented by its average model. The unit constitutes the average model of PDM inverter (**Figure 8**). The current drawn by this converter is as follows:

$$i_{dc}(t) = \frac{u_{dc}}{R_{eq}} \left\{ 1 - \frac{1 - exp\left(-(1-d) T_r/\tau_{eq} \right)}{1 - exp(-T_r/\tau_{eq})} exp(-t/\tau_{eq}) \right\} \text{ for } 0 \le t \le dT_{PDM} \tag{26}$$

$$i_{dc}(t) = 0 \text{ for } dT_{PDM} \langle t \le T_{PDM} \tag{27}$$

Using Eq. (26) and Eq. (27), the mean value, RMS value and form factor are calculated. We obtain:

$$\langle i \rangle_{dcT_{PDM}} = \frac{u_{dc}}{R_{eq}} \left\{ d - \frac{\tau_{eq}}{T_{PDM}} \frac{\left(1 - exp\left(-d\frac{T_{PDM}}{\tau_{eq}}\right)\right)\left(1 - exp\left(-(1-d)\frac{T_{PDM}}{\tau_{eq}}\right)\right)}{\left(1 - exp\left(-\frac{T_{PDM}}{\tau_{eq}}\right)\right)} \right\} \tag{28}$$

$$\langle I_{dc} \rangle = \frac{u_{dc}}{R_{eq}} \sqrt{d - 2\frac{\tau_{eq}}{T_{PDM}} \frac{\left(1 - e^{-d\frac{T_{PDM}}{\tau_{eq}}}\right)\left(1 - e^{-(1-d)\frac{T_{PDM}}{\tau_{eq}}}\right)}{1 - e^{-\frac{T_{PDM}}{\tau_{eq}}}} + \frac{\tau_{eq}}{2 \cdot T_{PDM}} \left(\frac{1 - e^{-(1-d)\frac{T_{PDM}}{\tau_{eq}}}}{1 - e^{-\frac{T_{PDM}}{\tau_{eq}}}} \right)^2 \left(1 - e^{-2d\frac{T_{PDM}}{\tau_{eq}}}\right)} \tag{29}$$

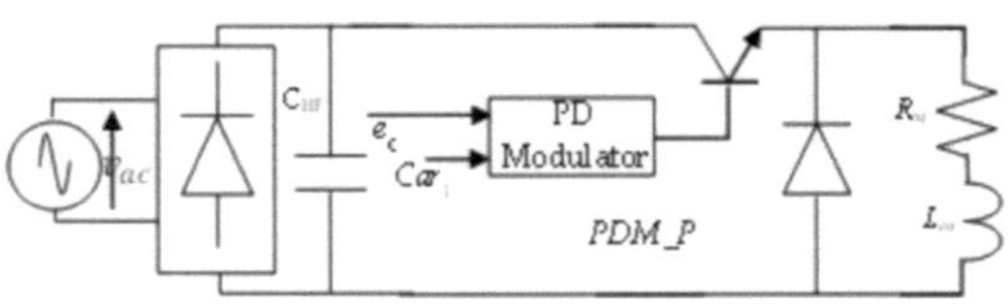

Figure 8. Average model of PDM inverter.

$$FF = \frac{d - \frac{\tau_{eq}}{T_{PDM}} \frac{\left(1-exp\left(-d\frac{T_{PDM}}{\tau_{eq}}\right)\right)\left(1-exp\left(-(1-d)\frac{T_{PDM}}{\tau_{eq}}\right)\right)}{1-exp\left(-\frac{T_{PDM}}{\tau_{eq}}\right)}}{\sqrt{d - 2\frac{\tau_{eq}}{T_{PDM}} \frac{\left(1-exp\left(-d\frac{T_{PDM}}{\tau_{eq}}\right)\right)\left(1-exp\left(-(1-d)\frac{T_{PDM}}{\tau_{eq}}\right)\right)}{1-exp\left(-\frac{T_{PDM}}{\tau_{eq}}\right)} + \frac{\tau_{eq}}{2 \cdot T_{PDM}} \left(\frac{1-exp\left(-(1-d)\frac{T_{PDM}}{\tau_{eq}}\right)}{1-exp\left(-\frac{T_{PDM}}{\tau_{eq}}\right)}\right)^2 \left(1 - exp\left(-2d\frac{T_{PDM}}{\tau_{eq}}\right)\right)}} \tag{30}$$

5.1.3. Extension to ac-ac converter

Now, we consider that the voltage u_{dc} is supplied by a single-phase diode rectifier. During the jth PDM pattern period, we suppose that this voltage is as follows:

$$u_{dc}(j) = \hat{V}_{ac} sin(\theta_j) \tag{31}$$

with $\theta_j = \frac{\pi}{q}\frac{2j-1}{2}$ is the midpoint of jth pattern period.

By substitution of Eq. (31) into Eqs. (28) and (29), we obtain mean value, RMS value and form factor of i_{dc} during jth pattern period:

$$\langle i_{dc} \rangle_{T_{PDM}} = \frac{\hat{V}_{ac}}{R_{eq}} \left\{ d - \frac{\tau_{eq}}{T_{PDM}} \frac{\left(1 - exp\left(-d\frac{T_{PDM}}{\tau_{eq}}\right)\right)\left(1 - exp\left(-(1-d)\frac{T_{PDM}}{\tau_{eq}}\right)\right)}{1 - exp\left(-\frac{T_{PDM}}{\tau_{eq}}\right)} \right\} sin(\theta_j) \tag{32}$$

$$I_{dc} = \frac{\hat{V}_{ac}}{R_{eq}} \sqrt{d - \frac{2\tau_{eq}}{T_{PDM}} \frac{1 - e^{-(1-d)\frac{T_{PDM}}{\tau_{eq}}}}{1 - e^{-\frac{T_{PDM}}{\tau_{eq}}}} \left\{ 1 - e^{-d\frac{T_{PDM}}{\tau_{eq}}} + \frac{1}{4}\left(\frac{1 - e^{-(1-d)\frac{T_{PDM}}{\tau_{eq}}}}{1 - e^{-\frac{T_{PDM}}{\tau_{eq}}}}\right)\left(1 - e^{-2d\frac{T_{PDM}}{\tau_{eq}}}\right) \right\}} sin(\theta_j) \tag{33}$$

In the appendix, we show that the fundamental component and the RMS value of the current drawn from ac-supply (Rectifier input current) and the mean and RMS values of the inverter input current during the jth PDM pattern period are linked by the following relationships:

$$\frac{\hat{I}_f}{(\hat{V}_{ac}/R_{eq})} = \frac{\langle i_{dc} \rangle_{T_{PDM}}(j)}{(\hat{V}_{ac}/R_{eq}) sin(\theta_j)} \tag{34}$$

$$\frac{I_{ac}}{(\hat{V}_{ac}/R_{eq})} = \frac{I_{dc}(j)}{(\hat{V}_{ac}/R_{eq}) sin(\theta_j)} \frac{1}{\sqrt{2}} \tag{35}$$

Knowing that in the case of a single-phase diode rectifier, the displacement factor is unitary, and the power factor is assimilated to the distortion factor:

$$PF = \frac{\hat{I}_f}{\sqrt{2} \cdot I_{ac}} \tag{36}$$

Substitution of Eqs. (35), (34), (33) and (32) into Eq. (36) yields:

$$PF = \frac{e_c - q2F_{ac}\tau_{eq}\frac{\left(1-exp(-e_c/q2F_{ac}\tau_{eq})\right)\left(1-exp\left(-(1-e_c)/q2F_{ac}\tau_{eq}\right)\right)}{1-exp(-1/q2F_{ac}\tau_{eq})}}{\sqrt{e_c - q4F_{ac}\tau_{eq}\frac{\left(1-exp\left(-\frac{e_c}{q2F_{ac}\tau_{eq}}\right)\right)\left(1-exp\left(-\frac{(1-e_c)}{q2F_{ac}\tau_{eq}}\right)\right)}{1-exp\left(-\frac{1}{q2F_{ac}\tau_{eq}}\right)} + qF_{ac}\tau_{eq}\left(\frac{1-e^{-\frac{(1-e_c)}{q2F_{ac}\tau_{eq}}}}{1-e^{-\frac{1}{q2F_{ac}\tau_{eq}}}}\right)^2\left(1-e^{-2\frac{e_c}{q2F_{ac}\tau_{eq}}}\right)}} \tag{37}$$

The transmitted power is defined by:

$$P = \frac{1}{2}\hat{V}_{ac}\hat{I}_f \tag{38}$$

Substitution of Eqs. (32) and (34) into Eq. (38) gives the transmitted power in pu:

$$p = e_c - q2F_{ac}\tau_{eq}\frac{\left(1 - exp(-e_c/q2F_{ac}\tau_{eq})\right)\left(1 - exp\left(-(1-e_c)/q2F_{ac}\tau_{eq}\right)\right)}{1 - exp(-1/q2F_{ac}\tau_{eq})} \tag{39}$$

Its reference is as follows:

$$P_{ref} = \frac{1}{2}\hat{V}_{ac}{}^2/R_{eq} \tag{40}$$

The expressions (37) and (39) can be greatly simplified, if we consider the hypothesis:

$$\text{H1} : q2F_{ac}\tau_{eq} \ll 1 \tag{41}$$

They become:

$$PF \approx \sqrt{e_c} \tag{42}$$

$$p \approx e_c \tag{43}$$

To establish the maximum value that $q2F_{ac}\tau_{eq}$ can take without the approximations becoming imprecise, we calculate the relative errors in the most unfavorable case:

$$\frac{\Delta PF}{PF} = \frac{|PF(e_c) - \sqrt{e_c}|}{PF(e_c)} \tag{44}$$

$$\frac{\Delta p}{p} = \frac{|p(e_c) - e_c|}{p(e_c)} \tag{45}$$

It is checked that the most unfavorable case, that is, the errors are maximal, occurs when e_c is minimal. For different values of $\boldsymbol{e_{c,min}}$, we plot these errors versus $\boldsymbol{q2F_{ac}\tau_{eq}}$ (**Figure 9**). It is noted that the power error is not limited (it increase continuously). Eq. (43) gives values that can be truly erroneous if $\boldsymbol{q2F_{ac}\tau_{eq}}$ is not kept below a maximum value $(\boldsymbol{q2F_{ac}\tau_{eq}})_{\boldsymbol{max}}$. This means that in order to maintain a relative error less than a given limit, when ***ec*** varies between 1 and $\boldsymbol{e_{c,min}}$, the coefficient q must respect the constraint:

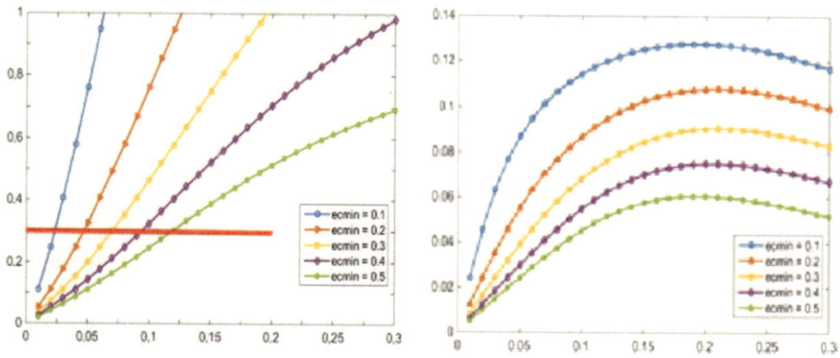

Figure 9. Power and power factor errors versus $q2F_{ac}\tau_{eq}$.

$$q \leq q_{max} = (q2F_{ac}\tau_{eq})_{max}/2F_{ac}\tau_{eq} \tag{46}$$

This constraint means that:

$$F_{car}\tau_{eq} \leq (q2F_{ac}\tau_{eq})_{max} \tag{47}$$

because

$$q = (F_{PDM} = F_{car})/2F_{ac} \tag{48}$$

In **Figure 9**, we can directly read $(q2F_{ac}\tau_{eq})_{max}$ as a function of the maximum permissible error and for different values of $e_{c,\,min}$. If, for example, to maintain the error below 30%, the carrier frequency must be less than the maximum frequency:

$$F_{car} \leq (F_{car})_{max} = 0.023/\tau_{eq} \text{ si } e_{c,\,min} = 0.1 \tag{49}$$

$$F_{car} \leq (F_{car})_{max} = 0.119/\tau_{eq} \text{ si } e_{c,\,min} = 0.5 \tag{50}$$

Eqs. (49) and (50) can be put in the general form:

$$F_{car}\tau_{eq} \leq \mathcal{L}_x(e_{c,\,min}) \tag{51}$$

where $\mathcal{L}_x(e_{c,\,min})$ is the maximum value that $F_{car}\tau_{eq}$ must not exceed if we want Eq. (43) to give the power with a tolerance less than x when e_c varies from 1 to $e_{c,\,min}$.

Eq. (26) shows that i_{dc} has a static component and a transient component:

$$i_{dc}{}^{s}(t) = \frac{u_{dc}}{R_{eq}} \text{ for } 0 \leq t \leq (dT_{PDM} = e_c/F_{car}) \tag{52}$$

$$i_{dc}{}^{t}(t) = \frac{u_{dc}}{R_{eq}} \frac{1 - exp\left(-(1-d)T_r/\tau_{eq}\right)}{1 - exp(-T_r/\tau_{eq})} exp(-t/\tau_{eq}) \text{ for } 0 \leq t \leq (dT_{PDM} = e_c/F_{car}) \tag{53}$$

Eq. (53) shows that the more we reduce $\mathcal{L}_x(e_{c,min})$, the more transient component is damped. Consequently, the currents become:

$$i_{dc}(t) = \frac{u_{dc}}{R_{eq}} * PDM\_P \tag{54}$$

$$i_{ac}(t) = \frac{u_{ac}}{R_{eq}} * PDM\_P \tag{55}$$

If the power and the power factor are determined from Eqs. (54) and (55), we find the expressions (42) and (43). This proves that the hypothesis H1, which allowed the passage of Eqs. (37) and (39) to Eqs. (42) and (43), indirectly signifies the predominance of the static component in i_{dc}.

5.2. Average model of ac-ac Multi-PDM converter (MPDMC)

The converter considered is that shown in **Figure 6**. By replacing each inverter by its average model, one builds the MPDMC's average model (**Figure 10**).

According to this average model, the current drawn by the gth inverter is as follows:

- zero, if its pattern is at zero:

$$i_{ac,g} = 0 \ if \ PDM\_P_g = 0 \tag{56}$$

- identical to the current drawn by a PDMC if its pattern is at 1, and the patterns of all the other inverters are at zero:

$$i_{ac,g} = i_{ac} \text{ if } PDM\_P_g = 1 \text{ and } PDM\_P_{j \neq g} = 0 \tag{57}$$

- identical to the current drawn by a PDMC multiplied by the number of inverters whose patterns are at 1, if its pattern is at 1:

$$i_{ac,g} = i_{ac} \sum_{j=1}^{G} PDM\_P_j \tag{58}$$

Eqs. (56), (57) and (58) can be written in the form:

$$i_{ac,g} = i_{ac} PDM\_P_g \sum_{j=1}^{G} PDM\_P_j \tag{59}$$

The current drawn by the MPDMC is the sum of the currents drawn by the various inverters:

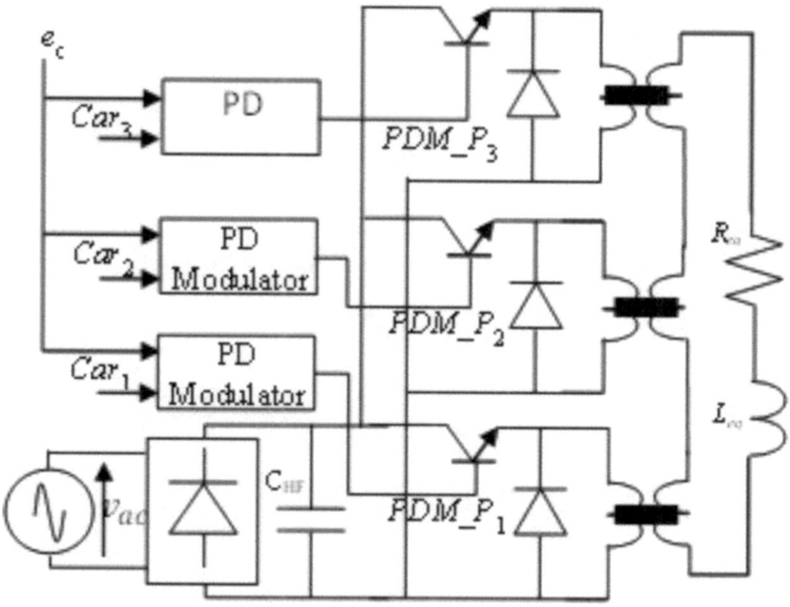

Figure 10. Average model of MPDMC.

$$i_{ac,MPDMC} = \sum_{g=1}^{G} i_{ac,g} \tag{60}$$

Substitution of Eqs. (59) and (60) into Eq. (55) yields:

$$i_{ac,MPDMC} = \frac{v_{ac}}{R_{eq}} \left(\sum_{g=1}^{G} PDM\_P_g \right)^2 \tag{61}$$

Comparing between Eqs. (61) and (55), we see that a M-PDMC is a PDMC that would be modulated by the square of the sum of the different patterns.

6. MPDMC features: power factor correction and transmitted power

The behavior of the MPDMC is determined by three elements: the control signal, the carrier and the coordination of all the carriers. In this section, we discuss three types of coordination.

6.1. First coordination: interlaced carriers

In **Figures 11** and **12**, we traced interlaced carriers, the patterns PDM_P_g and $\left(\sum_{g=1}^{G} PDM\_P_g\right)^2$ in the cases $G=2$ and $G=3$.

In the case $G=3$, when $0 \le e_c \le 1/3$, we have 1 or 0 pattern at 1:

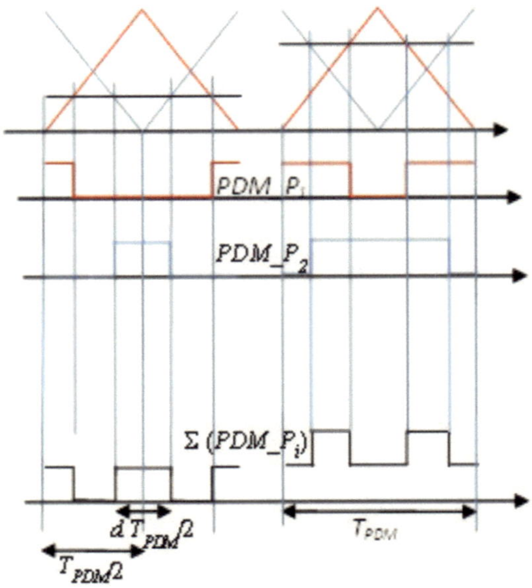

Figure 11. Interlaced carriers for $MPDMC_2$.

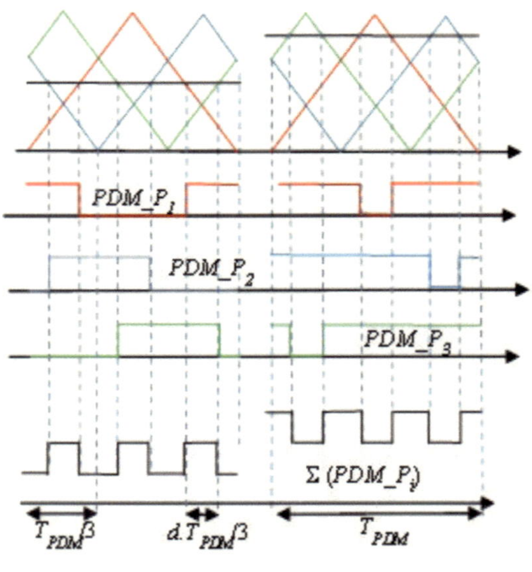

Figure 12. Interlaced carriers for $MPDMC_3$.

$$\left(\sum_{g=1}^{G} PDM\_P_g\right)^2 = 1 \text{ during } d(T_{PDM}/3) \tag{62.1.1}$$

$$\left(\sum_{g=1}^{G} PDM\_P_g\right)^2 = 0 \text{ during } (1-d)(T_{PDM}/3) \tag{62.1.2}$$

$$\text{with } d = 3.e_C \tag{62.1.3}$$

when $1/3 \le e_c \le 2/3$, we have 2 or 1 patterns at 1 :

$$\left(\sum_{g=1}^{G} PDM\_P_g\right)^2 = 2^2 \text{ during } d(T_{PDM}/3) \tag{62.2.1}$$

$$\left(\sum_{g=1}^{G} PDM\_P_g\right)^2 = 1^2 \text{ during } (1-d)(T_{PDM}/3) \tag{62.2.2}$$

$$\text{with } d = 3.e_C - 1 \tag{62.2.3}$$

when $2/3 \le e_c \le 3/3$, we have 3 or 2 patterns at 1 :

$$\left(\sum_{g=1}^{G} PDM\_P_g\right)^2 = 3^2 \text{ during } d(T_{PDM}/3) \tag{62.3.1}$$

$$\left(\sum_{g=1}^{G} PDM\_P_g\right)^2 = 2^2 \text{ during } (1-d)(T_{PDM}/3) \tag{62.3.2}$$

$$\text{with } d = 3.e_C - 2 \tag{62.3.3}$$

Eqs. (62) can be generalized as follows:

when $(g-1)/G \le e_c \le g/G$, we have g or $g-1$ patterns at 1 :

$$\left(\sum_{g=1}^{G} PDM\_P_g\right)^2 = g^2 \text{ during } d(T_{PDM}/G) \tag{63.1}$$

$$\left(\sum_{g=1}^{G} PDM\_P_g\right)^2 = (g-1)^2 \text{ during } (1-d)(T_{PDM}/G) \tag{63.2}$$

$$\text{with } d = G.e_C - (g-1) \tag{63.3}$$

$$\text{where } (g-1) = floor(G.e_C) \tag{63.4}$$

From Eq. (63), we determine average value and RMS value of $\left(\sum_{g=1}^{G} PDM\_P_g\right)^2$:

$$\left\langle \left(\sum_{g=1}^{G} PDM\_P_g\right)^2 \right\rangle = g^2 d + (g-1)^2(1-d) \tag{64}$$

$$RMS\left(\left(\sum_{g=1}^{G} PDM\_P_g\right)^2\right) = \sqrt{g^4 d + (g-1)^4(1-d)} \tag{65}$$

Using Eqs. (64) and (65), we determine transmitted power and power factor :

$$p = \left\langle \left(\sum_{g=1}^{G} PDM\_P_g \right)^2 \right\rangle = g^2 d + (g-1)^2 (1-d) \tag{66}$$

$$PF = \left\langle \left(\sum_{g=1}^{G} PDM\_P_g \right)^2 \right\rangle \Big/ RMS\left(\left(\sum_{g=1}^{G} PDM\_P_g \right)^2 \right) = \frac{1 + \left(\left(g/(g-1) \right)^2 - 1 \right) d}{\sqrt{1 + \left(\left(g/(g-1) \right)^4 - 1 \right) d}} \tag{67}$$

$$\text{with } d = G.e_C - (g-1) \tag{68}$$

$$\text{and } (g-1) = floor(G.e_C) \tag{69}$$

We find: (i) the same characteristic power versus control signal as in Section 5 and (ii) the power factor is determined by an analytic expression and not by a summation of the harmonics.

6.2. Second coordination: distribution in uniform bandwidths

Instead of reducing the total harmonic distortion by a mutual compensation of the individual distortions (produced by each inverter), one can reduce it differently: all inverters operate at full power or at standstill, and they do not produce distortion, except one that operates in modulation to ensure power variation. To achieve this correction, the carriers are distributed in uniform bandwidths. All carriers have the same peak value:

$$\widehat{Car}_g = 1/G \tag{70}$$

In **Figures 13** and **14**, we plot uniformly superimposed carriers, associated patterns PDM_P_g and $\left(\sum_{g=1}^{G} PDM\_P_g \right)^2$ in the cases $G=2$ and $G=3$.

In the case $G=3$, Eq. (70) leads to:

$$\widehat{Car}_1 = \widehat{Car}_2 = \widehat{Car}_3 = 1/3 \tag{71}$$

when $0 \le e_c \le 1/3$, we have 1 or 0 pattern at 1:

$$\left(\sum_{g=1}^{G} PDM\_P_g \right)^2 = 1 \text{ during } d.T_{PDM} \tag{72.1.1}$$

$$\left(\sum_{g=1}^{G} PDM\_P_g \right)^2 = 0 \text{ during } (1-d).T_{PDM} \tag{72.1.2}$$

$$\text{with } d = e_C / \widehat{Car}_1 = 3.e_C \tag{72.1.3}$$

when $1/3 \le e_c \le 2/3$, we have 2 or 1 patterns at 1:

http://dx.doi.org/10.5772/intechopen.68324

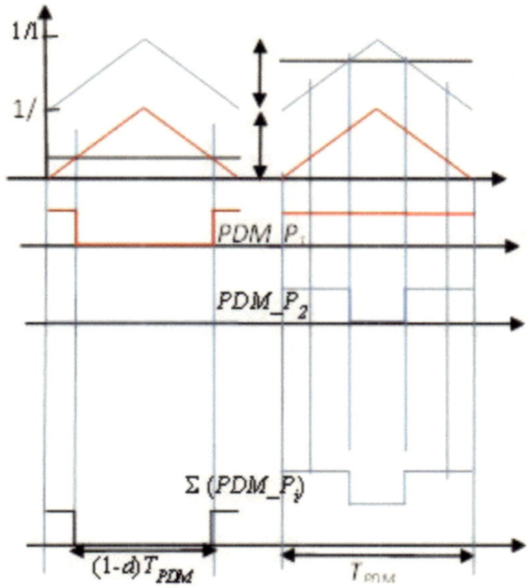

Figure 13. Carriers distributed in uniform bandwidths, $G=2$.

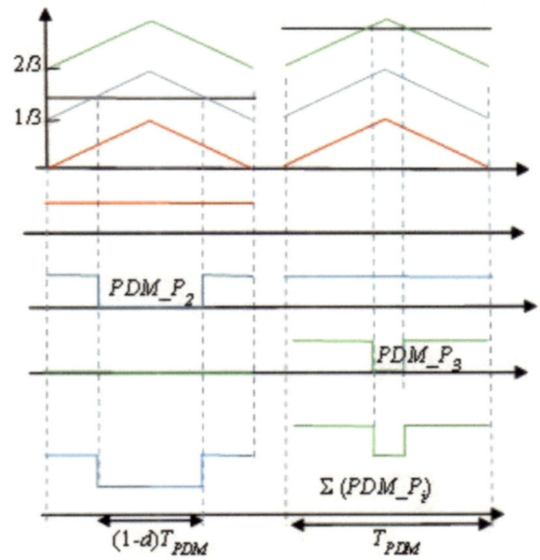

Figure 14. Carriers distributed in uniform bandwidths, $G=3$.

$$\left(\sum_{g=1}^{G} PDM\_P_g\right)^2 = 2^2 \text{ during } d.T_{PDM} \tag{72.2.1}$$

$$\left(\sum_{g=1}^{G} PDM\_P_g\right)^2 = 1^2 \text{ during } (1-d).T_{PDM} \tag{72.2.2}$$

$$\text{with } d = (e_C - 1/3)/\widehat{Car}_2 = 3.e_C - 1 \tag{72.2.3}$$

when $2/3 \leq e_c \leq 3/3$, we have 3 or 2 patterns at 1:

$$\left(\sum_{g=1}^{G} PDM\_P_g\right)^2 = 3^2 \text{ during } d.T_{PDM} \tag{72.3.1}$$

$$\left(\sum_{g=1}^{G} PDM\_P_g\right)^2 = 2^2 \text{ during } (1-d).T_{PDM} \tag{72.3.2}$$

$$\text{with } d = (e_C - 2/3)/\widehat{Car}_3 = 3.e_C - 2 \tag{72.3.3}$$

Eqs. (72) can be generalized as follows:

when $(g-1)/G \leq e_c \leq g/G$, we have g or $g-1$ patterns at 1:

$$\left(\sum_{g=1}^{G} PDM\_P_g\right)^2 = g^2 \text{ during } d.T_{PDM} \tag{73.1}$$

$$\left(\sum_{g=1}^{G} PDM\_P_g\right)^2 = (g-1)^2 \text{ during } (1-d).T_{PDM} \tag{73.2}$$

$$\text{with } d = \Big(G.e_C - (g-1)\Big)/(G.\widehat{Car}_g) \tag{73.3}$$

$$\textit{where } (g-1) = floor(G.e_C) \tag{73.4}$$

Using Eq. (73), we determine transmitted power and power factor:

$$p = \left\langle \left(\sum_{g=1}^{G} PDM\_P_g\right)\right\rangle^2 = g^2 d + (g-1)^2(1-d) \tag{74}$$

$$PF = \left\langle \left(\sum_{g=1}^{G} PDM\_P_g\right)\right\rangle^2 \Big/ RMS\left(\left(\sum_{g=1}^{G} PDM\_P_g\right)^2\right) = \frac{1 + \left(\left(g/(g-1)\right)^2 - 1\right)d}{\sqrt{1 + \left(\left(g/(g-1)\right)^4 - 1\right)d}} \tag{75}$$

$$\text{with } d = \Big(G.e_C - (g-1)\Big)/(G.\widehat{Car}_g) = G.e_C - (g-1) \tag{76}$$

$$\text{and } (g-1) = floor(G.e_C) \tag{77}$$

It is because the carriers are distributed in uniform bandwidths that power and power factor versus control signal characteristics are the same as in the previous case. The novelty is that the power characteristic is defined by one and only one carrier depending on the value of e_c. This is an advantage which facilitates the search of the conditions to correct the non-linearity of power characteristic.

6.3. Third coordination: distribution in non-uniform bandwidths

Using Eqs. (74) and (76), we determine the slope of the power characteristic

$$\frac{dp}{de_C} = \frac{g^2 - (g-1)^2}{\widehat{Car}_g} \tag{78}$$

To linearize this power characteristic, all segments of the power characteristic must have the same slope G^2 (because, when e_C varies from 0 to 1, power varies from 0 to G^2):

$$\frac{dp}{de_C} = \frac{g^2 - (g-1)^2}{\widehat{Car}_g} = G^2 \tag{79}$$

It is thus deduced that to linearize the power characteristic, it is necessary that:

$$\widehat{Car}_g = \frac{g^2 - (g-1)^2}{G^2} \tag{80}$$

The carriers are therefore distributed in non-uniform bandwidths defined by:

$$\left(BL_{g-1} = \sum_{j=1}^{g-1} \widehat{Car}_j\right) \leq e_C \leq \left(BL_g = \sum_{j=1}^{g} \widehat{Car}_j\right) \tag{81}$$

Taking account of Eq. (80), the lower and upper limits of a bandwidth become:

$$\begin{cases} BL_{g-1} = \sum_{j=1}^{g-1} \widehat{Car}_j = (g-1)^2/G^2 \\ BL_g = \sum_{j=1}^{g} \widehat{Car}_j = g^2/G^2 \end{cases} \tag{82}$$

For example, if one considers an MPDMC with three inverters, the three carriers and the three bandwidths are (**Figure 15**):

Car_1 : covers [0 1/9], its pic-value is $\widehat{Car}_1 = 1/9$ and its upper limit: $B_1 = 1/9$

Car_2 : covers [1/9 4/9], its pic-value is $\widehat{Car}_2 = 3/9$ and its upper limit: $B_2 = 4/9$

Car_3 : covers [4/9 9/9], its pic-value is $\widehat{Car}_3 = 5/9$ and its upper limit: $B_3 = 9/9$

The pattern duty cycle is defined by:

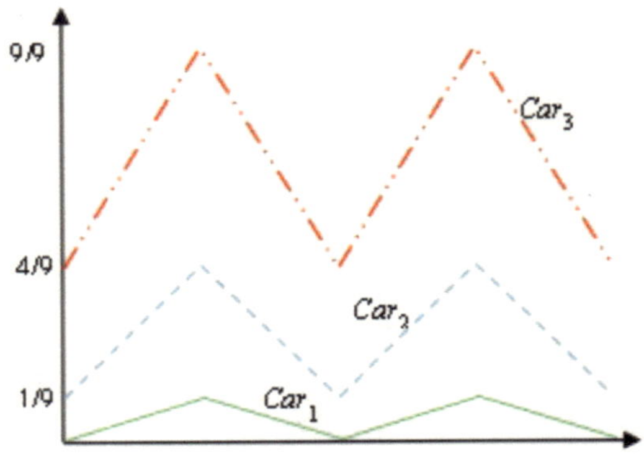

Figure 15. Carriers distributed in non-uniform bandwidths, G=3.

$$d = \frac{e_c - BL_{g-1}}{BL_g - BL_{g-1}} \text{ if } BL_{g-1} \leq e_c \leq BL_g \tag{83}$$

Substitution of Eq. (82) into Eq. (83) yields:

$$d = \frac{G^2 e_c - (g-1)^2}{2g-1} \text{ with } (g-1) = floor(G\sqrt{e_c}) \tag{84}$$

Substitution of Eq. (84) into Eq. (74) leads to the expression of transmitted power:

$$p = G^2 e_c \tag{85}$$

Substitution of Eq. (84) into Eq. (75) leads to the expression of transmitted power:

$$PF = \frac{G^2 e_c}{\sqrt{G^2 e_c \left(g^2 + (g-1)^2\right) - g^2 (g-1)^2}} \text{ with } (g-1) = floor(G\sqrt{e_c}) \tag{86}$$

Eqs. (85) and (86) show that the power characteristic is linear and that the power factor is unitary in G points:

$$PF = 1 \text{ when } e_c = BL_g = (g/G)^2 \text{ or when } p = g^2 \text{ with } g = 1, 2, \ldots G \tag{87}$$

7. Simulation results

Simulations are carried out in the Matlab SimPowerSystems environment. We consider a RLC load (1.85 Ω, 20 μH, 90 nF), matching transformer ration 3/10 and an ac-supply 120V-60Hz. The results of four simulation series are presented. **Figures 16** and **17** show examples of currents drawn by a PDMC and $MPDMC_2$. **Figures 18** and **19** show the theoretical and simulation results of transmitted power and power factor characteristics in the case of PDMC. **Figures 20** and **21** show the theoretical and simulation results of transmitted power and power factor characteristics in the case of $MPDMC_2$ and coordination's types 1, 2 and 3. We note a good agreement between theoretical and simulation results.

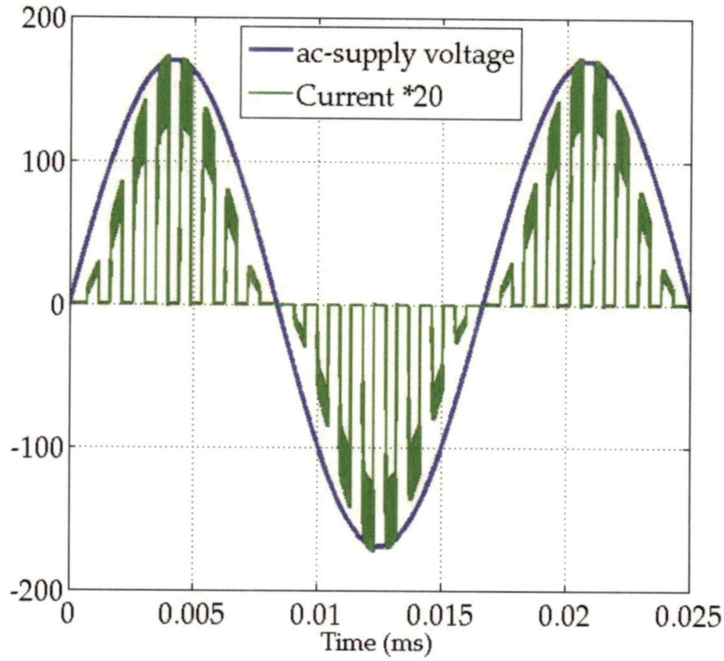

Figure 16. Current drawn by PDMC, $e_c = 0.5$.

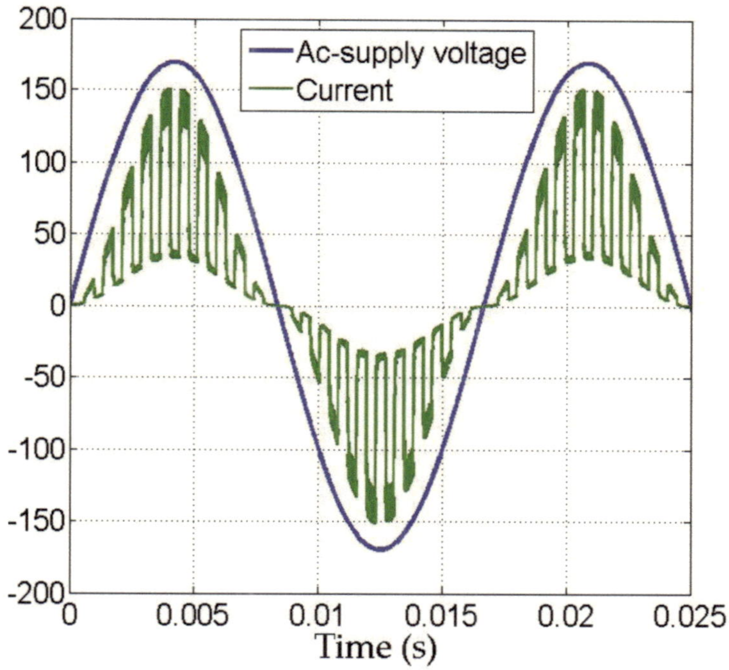

Figure 17. Current drawn by $MPDMC_2$ carriers coordination type 2, $e_c = 0.75$.

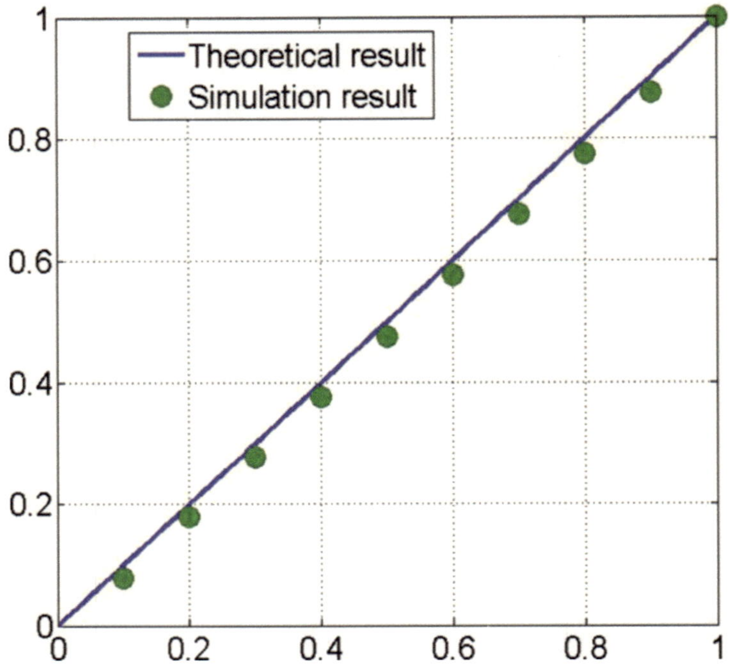

Figure 18. Power versus e_c characteristic of PDMC.

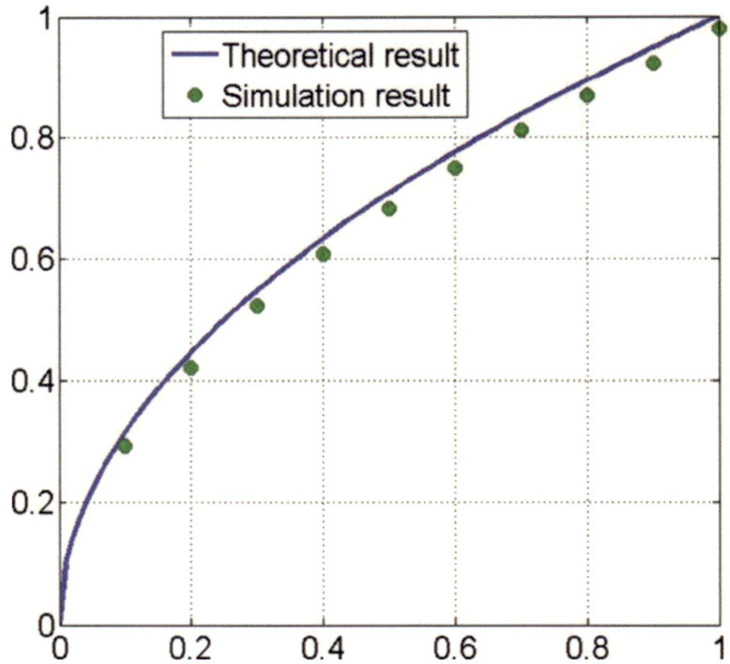

Figure 19. Power factor versus e_c characteristic of PDMC.

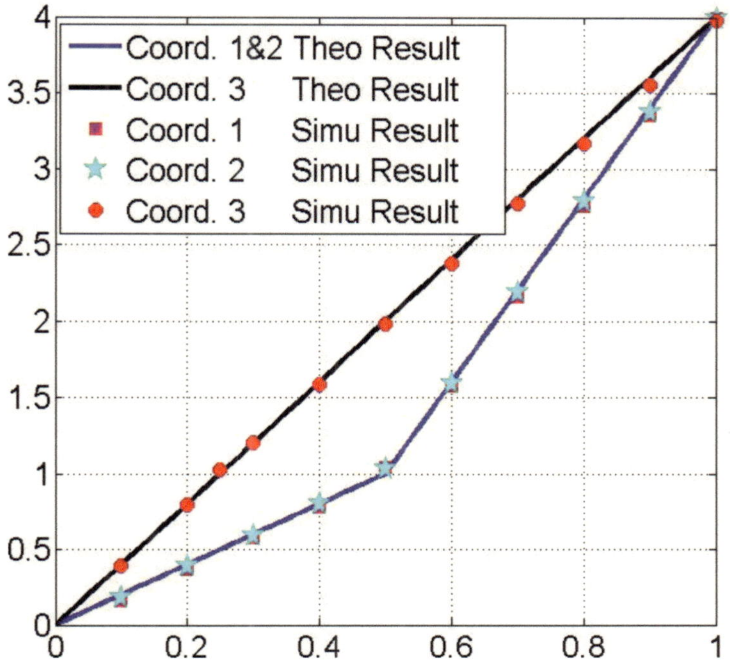

Figure 20. Power versus e_c characteristic of $MPDMC_2$.

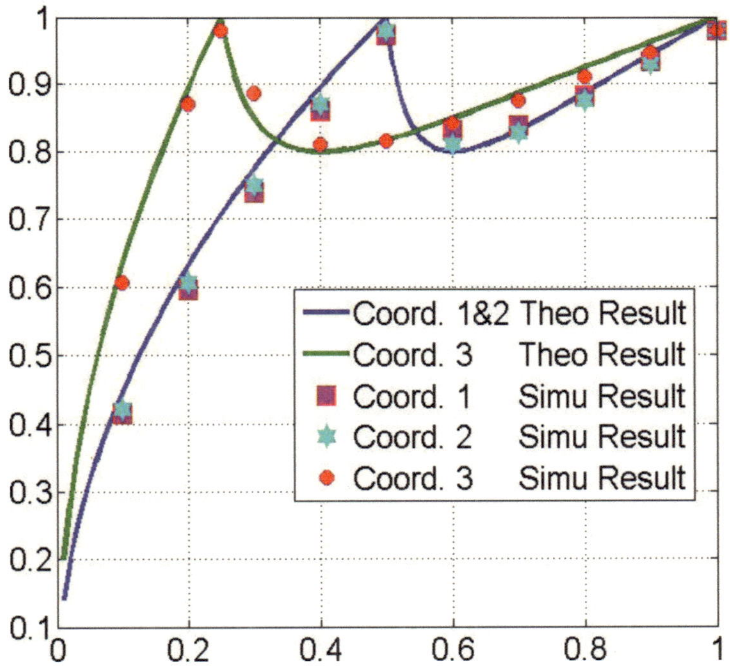

Figure 21. Power factor versus e_c characteristic of $MPDMC_2$.

8. Conclusion

The reflections and results presented in this chapter can be divided into two groups: the background and the novelties.

The background includes the PDM control principle, the PDM pattern generation and the characteristics of the ac-ac PDM converter and ac-ac Multi-PDM converter. In the description of the PDM control principle, we deliberately sought to present the PDM control as a form of association based on role sharing between the concepts of soft switching and hard switching. The role of soft switching is to produce lossless switching, while the role of hard switching is to vary the power by deciding the nature of the switching cycles. This reflection on the nature of the PDM control led us to represent the operation of the PDM inverter by a setting in cascade of a chopper and a series resonant inverter. We have detailed a PDM pattern generation method for the ac-ac PDM converter. It is a method inspired by PWM techniques, but adapted to the specificities of the PDM control (adaptation of the pattern frequency to both the resonant frequency and the ac-supply frequency). This method is based on a synchronous comparison of a carrier with a control signal and a calibration of the useful period of the dc voltage. Without the detailed, we have given the power transmitted characteristics and the spectrum of the current drawn from the ac-supply by an ac-ac PDM converter. Without detailed, we gave the characteristics of the transmitted power and the spectrum of the current drawn from the electrical communication by a ac-ac PDM converter. The power-factor correction by a total or partial mutual compensation is presented. It leads to the definition of a converter with several inverters and interlaced carriers. This is the ac-ac multi-PDM converter. The behavior of this converter is modeled by a system bi-converter.

The novelties include mainly the average modeling of ac-ac PDM converter and ac-ac multi-PDM converter and the introduction of carrier coordination as a control parameter of ac-ac multi-PDM converter. The application of average modeling leads to the representation of the series resonant inverter by an equivalent RL branch. The replacement of the inverter by its equivalent RL branch in the ac-ac PDM converter facilitates the analysis of this converter and makes it possible to establish (i) the conditions to preserve the linearity of the power characteristic and (ii) an analytical expression of the power-factor. The replacement of the inverter by its equivalent RL branch in the ac-ac Multi-PDM converter allows modeling this converter by an operating model (a model that integrates the operations of the components of the converter). This makes it possible to envisage several types of coordination. Three types of coordination are presented. Coordination by stratified carriers allows (i) a power-factor correction based on the search for a minimal distortion (ii) and to linearize the power characteristic.

Appendix

By subdividing $[0\ \pi]$ into q PDM pattern periods, the expressions of the fundamental component and the RMS-value of the drawn current are written:

http://dx.doi.org/10.5772/intechopen.68324

$$\hat{I}_{ac,f} = \frac{2}{\pi}\sum\nolimits_{j=1}^{q}\int_{(j-1)\pi/q}^{j\pi/q} i_{dc}sin(\theta)d\theta \tag{A.1}$$

$$I_{ac}{}^2 = \frac{2}{\pi}\sum\nolimits_{j=1}^{q}\int_{(j-1)\pi/q}^{j\pi/q} i_{dc}{}^2 d\theta \tag{A.2}$$

Assuming that q is large enough so that the sin varies very little over the interval $[(j-1)\pi/q \; j\pi/q]$, we write that:

$$sin(\theta) = sin\Big(\theta_j = (\pi/q)(j-1/2)\Big) \tag{A.3}$$

Taking into account Eqs. (A.3) and (A.1) becomes:

$$\hat{I}_{ac,f} = \frac{2}{\pi}\sum\nolimits_{j=1}^{q} sin(\theta_j)\int_{(j-1)\pi/q}^{j\pi/q} i_{dc}d\theta \tag{A.4}$$

Knowing that average and RMS values of i_{dc} are as follows:

$$\langle i_{dc}\rangle(j) = \frac{\pi}{q}\int_{(j-1)\pi/q}^{j\pi/q} i_{dc}d\theta \tag{A.5}$$

$$\boldsymbol{I_{dc}(j)^2 = \frac{q}{\pi}\int_{(j-1)\pi/q}^{j\pi/q} i_{dc}{}^2 d\theta} \tag{A.6}$$

It is established that:

$$\hat{I}_{ac,f} = \frac{2}{q}\sum_{j=1}^{q} sin(\theta_j)\langle i_{dc}\rangle(j) \tag{A.7}$$

$$\boldsymbol{I_{ac}{}^2 = \frac{1}{\pi}\sum_{j=1}^{q}\frac{\pi}{q} I_{dc}(j)^2} \tag{A.8}$$

Substitutions of Eq. (32) into Eqs. (A.7) and (37) into Eq. (A.8) yield:

$$\hat{I}_{ac,f} = \frac{\hat{V}_{ac}}{R_{eq}}\left\{d - \frac{\tau_{eq}}{T_{PDM}}\frac{\left(1-exp\left(-d\frac{T_{PDM}}{\tau_{eq}}\right)\right)\left(1-exp\left(-(1-d)\frac{T_{PDM}}{\tau_{eq}}\right)\right)}{1-exp\left(-\frac{T_{PDM}}{\tau_{eq}}\right)}\right\}\frac{2}{q}\sum\nolimits_{j=1}^{q} sin(\theta_j)^2 \tag{A.9}$$

$$\frac{I_{ac}{}^2}{\left(\frac{\hat{V}_{ac}}{R_{eq}}\right)^2} = \left\{d - \frac{2\tau_{eq}}{T_{PDM}}\frac{1-e^{-(1-d)\frac{T_{PDM}}{\tau_{eq}}}}{1-e^{-\frac{T_{PDM}}{\tau_{eq}}}}\left\{1-e^{-d\frac{T_{PDM}}{\tau_{eq}}} + \frac{\left(1-e^{-(1-d)\frac{T_{PDM}}{\tau_{eq}}}\right)\left(1-e^{-2d\frac{T_{PDM}}{\tau_{eq}}}\right)}{4\left(1-e^{-\frac{T_{PDM}}{\tau_{eq}}}\right)}\right\}\right\}\frac{\sum\nolimits_{j=1}^{q} sin(\theta_j)^2}{q} \tag{A.10}$$

Knowing that:

$$\sum_{j=1}^{q} sin(\theta_j)^2 = q/2 \tag{A.11}$$

Eqs. (A.9) and (A.10) becomes:

$$\hat{I}_{ac,f} = \frac{\hat{V}_{ac}}{R_{eq}}\left\{d - \frac{\tau_{eq}}{T_{PDM}}\left(1 - exp\left(-d\frac{T_{PDM}}{\tau_{eq}}\right)\right)\left(1 - exp\left(-(1-d)\frac{T_{PDM}}{\tau_{eq}}\right)\right)/1 - exp\left(-\frac{T_{PDM}}{\tau_{eq}}\right)\right\} \tag{A.12}$$

$$I_{ac}{}^2 = \frac{\hat{V}_{ac}{}^2}{2R_{eq}{}^2}\left\{d - \frac{2\tau_{eq}}{T_{PDM}}\frac{1 - e^{-(1-d)\frac{T_{PDM}}{\tau_{eq}}}}{1 - e^{-\frac{T_{PDM}}{\tau_{eq}}}}\left\{1 - e^{-d\frac{T_{PDM}}{\tau_{eq}}} + \frac{\left(1 - exp\left(-d\frac{T_{PDM}}{\tau_{eq}}\right)\right)\left(1 - exp\left(-(1-d)\frac{T_{PDM}}{\tau_{eq}}\right)\right)}{4\left(1 - exp\left(-\frac{T_{PDM}}{\tau_{eq}}\right)\right)}\right\}\right\} \tag{A.13}$$

Taking into account Eqs. (32) and (33), we obtain:

$$\hat{\boldsymbol{I}}_{ac,f} = \langle \boldsymbol{i}_{dc} \rangle_{T_{PDM}}(\boldsymbol{j})/\boldsymbol{sin}(\boldsymbol{\theta}_j) \tag{A.14}$$

$$\boldsymbol{I}_{ac} = \boldsymbol{I}_{dc}(\boldsymbol{j})/\sqrt{2}\boldsymbol{sin}(\boldsymbol{\theta}_j) \tag{A.15}$$

Nomenclature

$A\left(= exp(-\alpha T_r/2)\right)$	Intermediate constant
C	Load's capacitor
F_{ac} (T_{ac})	Ac-supply frequency (period)
L	Load's inductor
R	Load's resistor
T_{dc} ($= T_{ac}/2$)	Period of voltage rectified
T_{PDM}	PDM pattern
T_r	Resonance period
$\hat{V}_{ac}$	Amplitude of ac-supply voltage
d	Duty cycle of PDM pattern
$floor$	Integer portion
k	PDM pattern length
q	PDM pattern frequency in pu ($F_{PDM} = q2F_{ac}$)
$\alpha(= R/2L)$	Attenuation factor
$\tau_{eq}(= L_{eq}/R_{eq})$	Time constant of RL equivalent branch
$\xi\left(= (R/2)\sqrt{C/L}\right)$	Damping ratio

Author details

Abdelhalim Sandali[1]* and Ahmed Chériti[2]

*Address all correspondence to: sandali@uqtr.ca

1 University Hassan II Casablanca - ENSEM, Casablanca, Morocco

2 University of Quebec at Trois-Rivières, Trois-Rivières, Quebec, Canada

References

[1] Divan DM, Venkataramanan G, De Doncker RW. Design methodologies for soft switched inverters. In: IEEE, editor. IAS Annual meeting; October 2–7; Pittsburgh. Pennsylvania: IEEE; 1988. pp. 758–766. DOI: 10.1109/IAS.1988.25147

[2] Liu KH, Oruganti R, Lee FC. Quasi-resonant converters topologies and characteristics. IEEE-Transactions on Power Electronics. 1987;**PE-2**(1):62–71. DOI: 10.1109/TPEL.1987.4766333

[3] Li R, Xu D. A zero-voltage switching three-phase inverter. IEEE Transactions on Power Electronics. 2014;**29**(3):1200–1210. DOI: 10.1109/TPEL.2013.2260871

[4] Outeiro MT, Buja G, Czarkowski D. Resonant power converters: an overview with multiple elements in the resonant tank network. IEEE Industrial Electronics Magazine. 2016;**10**(2):21–45. DOI: 10.1109/MIE.2016.2549981

[5] Enhui Chu, Xing Zhang, Liang Huang. Research on a novel modulation strategy fot double auxiliary resonant commuted pole soft-switching inverter with the shunt dead time. IEEE Transactions on Power Electronics. 2016;**31**(10):6855–6869. DOI: 10.1109/TPEL.2015.2508143

[6] Fujita H, Akagi H. Pulse density modulated power control of a 4 kW, 450 kHz voltage-source inverter for induction melting applications. IEEE Transactions on Industry Applications. 1996;**32**(2):279–286. DOI: 10.1109/28.491475

[7] Fujita H, Akagi H. Control and performance of a pulse density modulated series-resonant inverter for corona discharge processes. IEEE Transactions on Industry Applications. 1999;**35**(3):621–227. DOI: 10.1109/28.767013

[8] Esteve V, Jordán J, Sanchis-Kilders E, Dede Enrique J, Maset E, Ejea Juan B, Ferreres A. Enhanced pulse-density-modulated power control for high-frequency induction heating inverters. IEEE Transactions on Industry Electronics. 2015;**62**(11):6905–6914. DOI: 10.1109/TIE.2015.2436352

[9] Lucia O, Sarnago H, Burdio JM. Pulse density modulated control for the series resonant multi-inverter for induction heating applications. In: IEEE, editor. IECON—42nd Annual

Conference of the IEEE Industrial Electronics Society; October 24–27; Firenze (Florence), Italy: 2016. pp. 5995–6000. DOI: 10.1109/IECON.2016.7794026

[10] Mousavian H, Abnavi S, Bakhshai A, Jain P. A push-pull Class E converter with improved PDM control. In: IEEE, editor. 7th International Symposium on Power Electronics for Distributed Generation Systems (PEDG); 27–30 June; Vancouver, Canada. Vancouver, Canada: IEEE; 2016. DOI: 10.1109/PEDG.2016.7527053

[11] Mishima T, Nakaoka M. A load-power adaptive dual pulse modulated current phasor-controlled ZVS high-frequency resonant inverter for induction heating applications. IEEE Transactions on Power Electronics. 2014;**29**(8):3864–3880. DOI: 10.1109/TIE.2013.2288985

[12] Sandali A, Cheriti A, Sicard P. Design considerations for PDM Ac-ac converter implementation. In: IEEE, editor. Annual IEEE Applied Power Electronics Conference and Exposition (APEC'07); February 25–March 1; Anaheim, California: IEEE; 2007. pp. 1678–1683. DOI: 10.1109/APEX.2007.357744

[13] Sandali A, Cheriti A, Sicard P, Al Haddad K. Application of PDM control to a multilevel ac-ac converter with self power factor correction. In: IEEE, editor. 35th Annual Power Electronics Specialists Conference (PESC'04); June 20–25; Aachen, Germany: IEEE; 2004. pp. 2881–2887. DOI: 10.1109/PESC.2004.1355291

[14] Lin W, Jovcic D. Average modelling of medium frequency DC-DC converters in dynamic studies. IEEE Transactions on Power Delivery. 2015;**30**(1):281–289. DOI: 10.1109/TPWRD.2014.2321425

[15] Xiao Liu, Aaron M. Cramer, Fei Pan. Generalized average method for time-invaring modeling of inverters. IEEE Transactions on Circuits and Systems. 2017;**64**(3):740–751. DOI: 10.1109/TCSI.2016.2620442

[16] Sandali A, Cheriti A. Accurate average model of PDM dual inverter. International Review on Modelling and Simulations. 2014;**7**(5):816–828. DOI: ISSN-19749821

Chapter 5

Dual-Inverter Circuit Topologies for Supplying Open-Ended Loads

Javier Riedemann Aros, Rubén Peña Guíñez and Ramón Blasco Gimenez

Additional information is available at the end of the chapter

http://dx.doi.org/10.5772/intechopen.68450

Abstract

Power electronic converters are nowadays the most suitable solution to provide a variable voltage/current in industry. The most commonly used power converter is the three-phase two-level voltage source inverter which transforms a direct-current input voltage into alternating-current output voltage with adjustable magnitude and frequency. Power inverters are used to supply three-phase loads which are typically connected in wye or delta configurations. However, in previous years, a type of connection consisting on leaving both terminal ends of the load opened has been studied as an alternative to standard wye or delta connection. To supply loads with this type of connection, two power inverters (one at each terminal end of the load) are required in a circuit topology called dual-inverter. In this chapter, a general study of the dual-inverter topology is presented. The advantages and issues of such converter are studied and different modulation strategies are shown and discussed. Moreover, multilevel dual-inverter converters are presented as an extension to the basic two-level idea. For evaluation purposes, simulations results are presented.

Keywords: voltage source inverter, dual-inverter, open-end winding, pulse width modulation

1. Introduction

In industrial applications, typical loads generally require to be supplied with alternating voltage of variable magnitude and frequency. To produce such voltages, power electronic converters are nowadays the standard and more suitable solution. The most commonly used power converter is the three-phase two-level voltage source inverter (VSI) which transforms a DC input voltage into AC output voltage with adjustable magnitude and frequency (**Figure 1**) [1, 2]. Three-phase VSIs typically supply loads connected in delta (called closed connection) or in wye (called semi-closed connection) (**Figure 2**), depending on the load requirements of voltage and current.

Regarding the VSI, to produce an AC output voltage, a modulation scheme should be used. The carrier-based pulse width modulation (PWM) strategy is a standard modulation technique for power inverters [2] where a triangular (carrier) signal v_{tri} is compared with a sinusoidal (reference) signal v_{ref}, as shown in **Figure 3**. The following control logic is used to generate the VSI-IGBTs gate pulses:

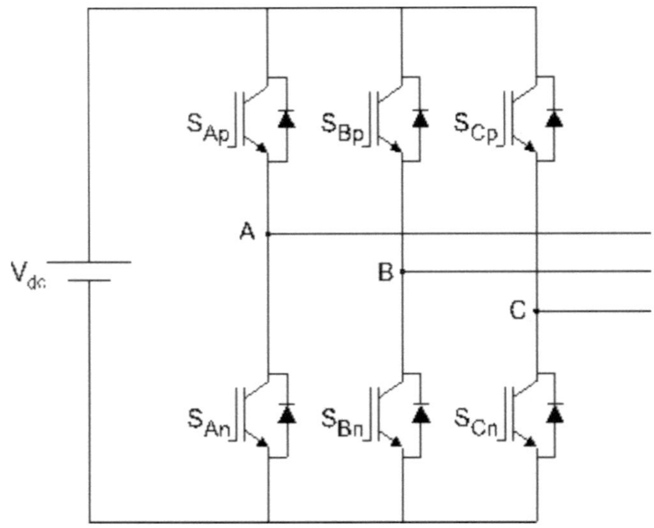

Figure 1. Two-level voltage source inverter.

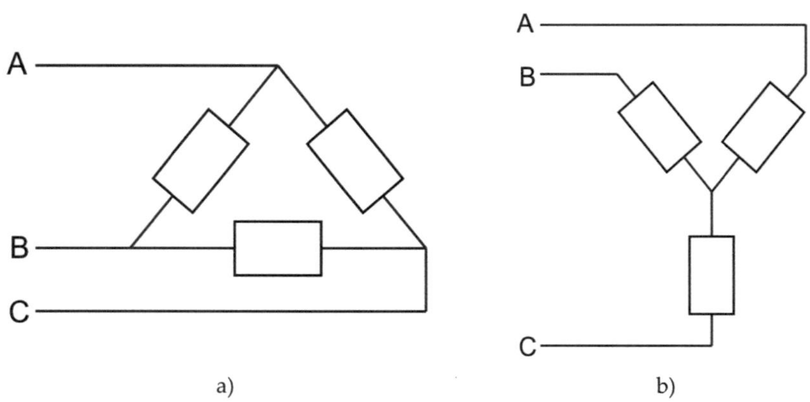

Figure 2. (a) Delta connection and (b) wye connection.

$$v_{ref} \geq v_{tri} \Rightarrow S_{xp} = 1$$
$$v_{ref} < v_{tri} \Rightarrow S_{xp} = 0$$

where S_{xp} with $x = A, B, C$ is an upper switch of the inverter.

To modulate the three legs of the inverter, three sinusoidal reference signals are required of equal magnitude and frequency but phase shift 120°.

The other standard PWM strategy for three-phase VSIs is the space vector modulation (SVM) where the possible switching states of the inverter are expressed as space vectors (**Table 1**) which when represented graphically form a hexagon divided in six sectors (**Figure 4**). The reference vector (v_{ref}) that represents the desired output voltage of the VSI should be synthetized using the available switching states in a sector [2].

To apply the switching states of the inverter, the following duty cycles are considered for the active vectors [1]:

$$d_\alpha = m \cdot \sin\left(\frac{\pi}{3} - \theta_{ref,o}\right), \; d_\beta = m \cdot \sin\left(\theta_{ref,o}\right) \tag{1}$$

where $\theta_{ref,o}$ is the angle of the output reference voltage vector and m is a modulation index.

The duty cycle of the zero vectors is given by

$$d_0 = 1 - d_\alpha - d_\beta \tag{2}$$

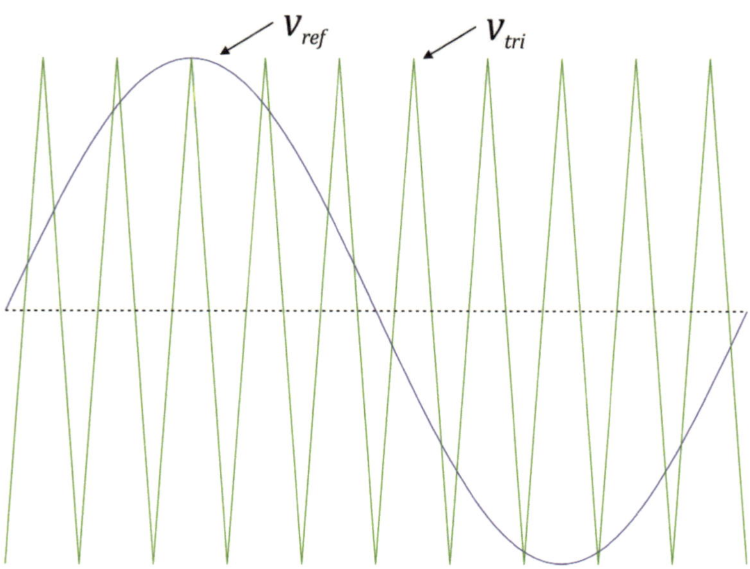

Figure 3. Signals used in a carrier-based modulation strategy (one leg of the inverter).

States of inverter $[S_A S_B S_C]$			
$V_1 = [100]$	$V_2 = [110]$	$V_3 = [010]$	$V_4 = [011]$
$V_5 = [001]$	$V_6 = [101]$	$V_7 = [111]$	$V_8 = [000]$

Table 1. Switching vectors of a VSI.

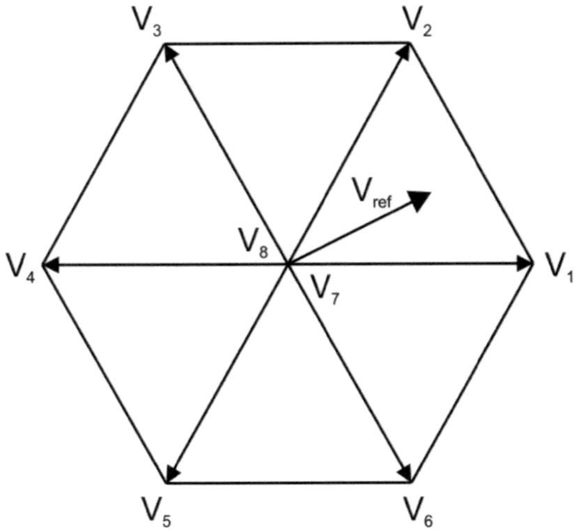

Figure 4. Graphic representation of the switching vectors.

A representation of the switching sequence in a switching period for each leg of the VSI is shown in **Figure 5**. As can be noted, the switching sequence aims to change the state of one switch at a time, then reducing the switching losses of the inverter.

Independent of the modulation strategy used, the standard two-level VSI supplying delta- or wye-connected loads has been widely studied for years and is a well-known and reliable engineering solution in the industry. However, in previous years, a type of connection consisting on leaving both terminal ends of the load opened has been studied as an alternative to standard wye or delta connection (**Figure 6**). To supply loads with this type of connection, two VSIs are required in a circuit topology called dual-inverter [1]. The dual-inverter circuit can be supplied by isolated DC sources (**Figure 7**) [3, 4] or by a single DC source (**Figure 8**) [5, 6]. It can be noted that when supplying the dual-inverter with a single DC source is equivalent to supplying each phase load with a single-phase VSI (H-bridge), hence the modulation scheme could be unipolar or bipolar [2].

Open-ended load connection offers certain advantages compared to wye or delta connections, such as [7, 8]:

- Equal power input from both sides of the load; thus, each VSI is rated at half the load power rating.
- Each load phase current can be controlled independently.

http://dx.doi.org/10.5772/intechopen.68450

- Possibility to have twice the effective switching frequency (depending on the modulation strategy).
- Possibility of reducing the common-mode voltage (CMV).
- Extensible to more phases, therefore multiphase loads can be considered if current reduction is required.

However, an open-ended load can have some drawbacks, such as:

- Possibility of zero sequence current flowing in the machine because of the occurrence of zero sequence voltage (ZSV).
- More complex power converter requirements, i.e. more power devices, circuit gate drives, etc.
- Complexity could affect the reliability.
- Greater weight and volume of the power converter.

When using two isolated DC sources to supply a dual-inverter (**Figure 7**), the main feature is that the circulation of zero sequence current is avoided due to the circuit configuration; therefore, the focus should be in reducing the common-mode voltage. Nevertheless, this topology requires two isolated DC sources which means two isolation transformers, increasing the cost and volume of the converter.

Several articles have been published with the use of this circuit configuration. For instance, Ref. [3] proposes a voltage harmonic suppression scheme for the dual-inverter, whereby selecting a

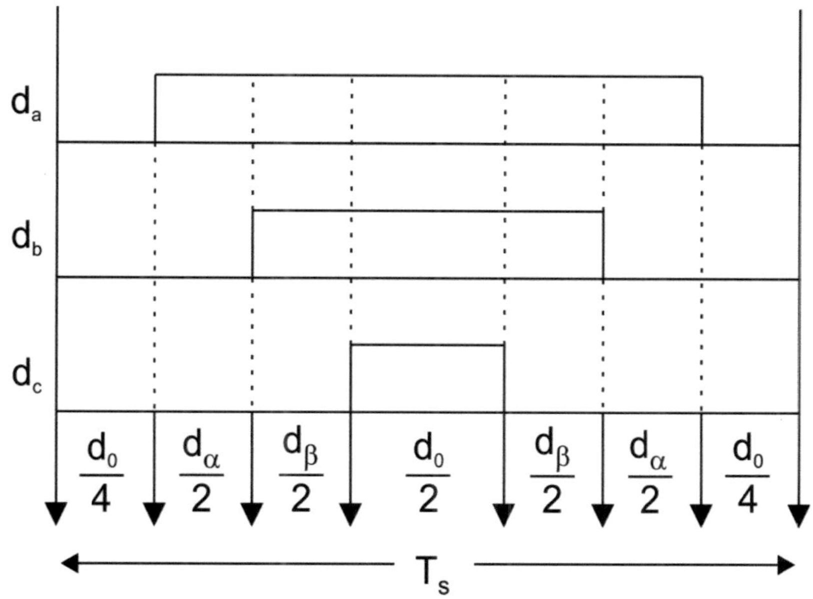

Figure 5. Switching sequence in one period.

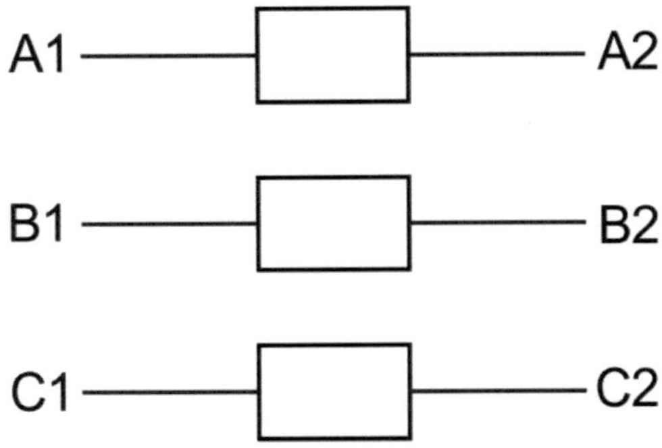

Figure 6. Open-ended load.

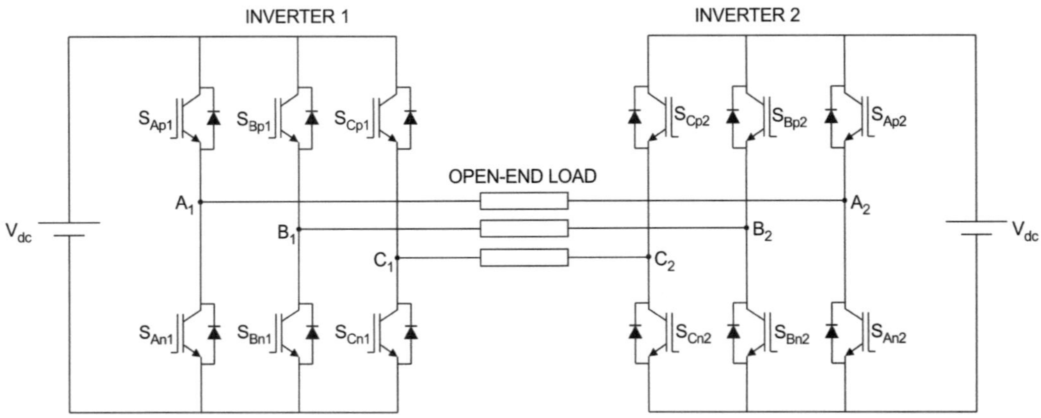

Figure 7. Two two-level VSIs fed by isolated DC sources for an open-ended load.

specific magnitude ratio of the DC sources, certain harmonic components of the output phase voltage can be eliminated. In Ref. [9], a method to extend the operating speed range of an open-end winding electrical machine is proposed based on the voltage range enhancement.

In Ref. [10], a modulation strategy for reducing the voltage total harmonic distortion (THD) in a dual-inverter is presented consisting on adjusting the pulses times of one of the inverters with respect to the other. In Ref. [11], a unified SVM strategy is proposed for a dual two-level inverter system in accordance with the voltage-second integral principle and the ratio of the two DC sources can be arbitrary positive values.

To reduce the switching losses of a dual-inverter, a space vector modulation (SVM) strategy is presented in Ref. [12]. The strategy does not require sector identification and allows a reduction of 50% in the switching losses, in comparison to other dual-inverter PWM techniques where one inverter is clamped at a determined switching state, while the other inverter commutates. In Ref. [13], a comparative study between three different modulation strategies for open-ended windings AC machine drives is carried out. Simulation results are presented and the current ripple under each PWM method is analysed.

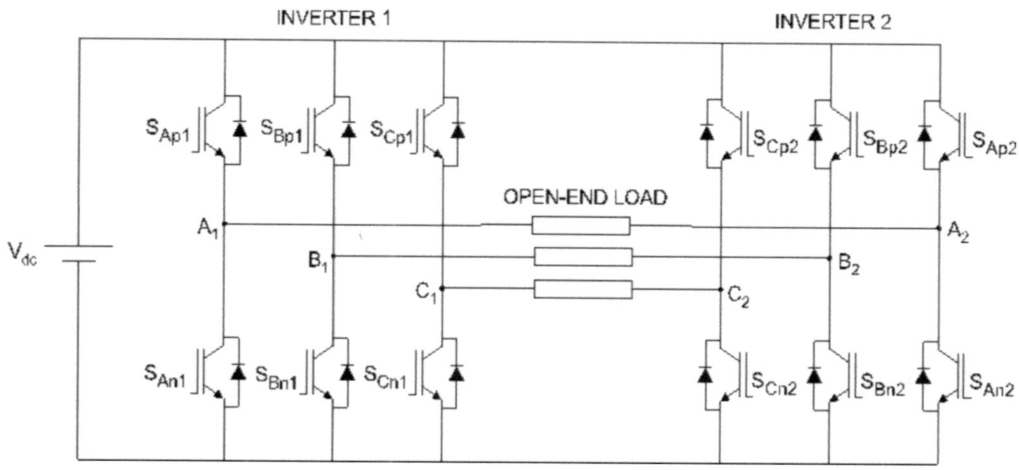

Figure 8. Two two-level VSIs fed by a single DC source for an open-end load.

The extension of the three-phase open-ended windings AC machine drive to multiphase machines is presented in Refs. [14–17] where different modulation strategies are presented and discussed. This extension is not straightforward because the number of possible switching states increases exponentially with the number of phases.

On the other hand, the dual-inverter supplied by a single DC source is a cheaper and of lower volume alternative, but circulation of zero sequence current could occur if zero sequence voltage is produced. Therefore, the attention should be put on reducing the common-mode voltage as well as the zero sequence voltage.

To eliminate the occurrence of zero sequence currents in the load, PWM strategies intended to eliminate the zero sequence voltage are proposed in Refs. [5, 6]. In Ref. [18], a SVM strategy is proposed to dynamically compensate the zero sequence current by applying the null vectors with asymmetrical duty cycles in each switching period and, in Ref. [19], the effect of the null-vector placement in the modulation for the dual-inverter system is thoroughly analysed. On the other hand, a closed-loop compensation scheme to suppress the zero sequence currents in the machine is developed in Ref. [20].

In Ref. [21], a vector control scheme for an open-end winding permanent magnet synchronous motor (PMSM) is presented considering a regulation mechanism for the zero sequence voltage, whereas in Ref. [22], a closed-loop control strategy intended to reduce the torque ripple in a PMSM with non-sinusoidal back electromotive force (EMF) is proposed.

To obtain a common-mode voltage reduction, a SVM switching strategy is presented in Ref. [23]. This modulation strategy considers only voltage space vectors that do not produce common-mode voltage then reducing the problems associated to it such as the bearing currents.

A dual-inverter configuration fed by an active rectifier without DC-link energy storage element (so-called direct-link converter) is presented in Refs. [24, 25]; in Ref. [24], three modulation strategies are presented for the drive: a carrier-based PWM and two SVM strategies. In Ref. [25], common-mode voltage suppression is proposed and an active filter is added to the topology to inject compensating harmonic currents into the supply and allow controllable input power factor.

Finally, multiphase open-ended windings induction motor drives are presented in Refs. [26–29], where the proposed PWM techniques are intended to reduce the common-mode voltage at the machine terminals; simulation and experimental results are also shown.

In this chapter, a general study of the dual-inverter topology is presented. The issues of zero sequence and common-mode voltages are thoroughly studied and different modulation strategies for the converter are shown and discussed; for evaluation purposes, simulations results are presented. Although the chapter is mainly focused on a two-level dual-inverter system supplied with isolated and non-isolated DC source, multilevel dual-inverter topologies are also discussed briefly.

2. The dual two-level inverter

In this section, a mathematical model of a two-level dual-inverter system will be developed. The model allows understanding the generation of the phase output voltage produced by the dual-inverter by means of the output voltage produced by the individual VSIs. The basic circuit configuration for supplying an open-ended load consists on connecting a standard two-level VSI at each side of the load (**Figures 7** and **8**). The output pole voltage of Inverter 1 ($\boldsymbol{v}_{o1}$) and Inverter 2 ($\boldsymbol{v}_{o2}$) with respect to the negative DC-link rail, is defined by

$$\boldsymbol{v}_{o1} = \boldsymbol{S}_{i1} \cdot v_{DC}, \ \boldsymbol{v}_{o2} = \boldsymbol{S}_{i2} \cdot v_{DC} \tag{3}$$

where the switching matrices of Inverter 1 ($\boldsymbol{S}_{i1}$) and Inverter 2 ($\boldsymbol{S}_{i2}$) are as follows:

$$\boldsymbol{S}_{i1} = \begin{bmatrix} S_{A1} \\ S_{B1} \\ S_{C1} \end{bmatrix} = \begin{bmatrix} S_{Ap1} - S_{An1} \\ S_{Bp1} - S_{Bn1} \\ S_{Cp1} - S_{Cn1} \end{bmatrix}, \ \boldsymbol{S}_{i2} = \begin{bmatrix} S_{A2} \\ S_{B2} \\ S_{C2} \end{bmatrix} = \begin{bmatrix} S_{Ap2} - S_{An2} \\ S_{Bp2} - S_{Bn2} \\ S_{Cp2} - S_{Cn2} \end{bmatrix}. \tag{4}$$

and $S_{xpx} = \overline{S}_{xnx} \in \{0, 1\}$ with $x = a, b, c, k = 1, \ 2$. The output phase voltages correspond to the pole voltage difference of both inverters:

$$\boldsymbol{v}_{ph,o} = \begin{bmatrix} v_{ph,oa} & v_{ph,ob} & v_{ph,oc} \end{bmatrix}^T = \boldsymbol{v}_{o1} - \boldsymbol{v}_{o2} = (\boldsymbol{S}_{i1} - \boldsymbol{S}_{i2}) v_{DC} \tag{5}$$

The three voltages produced by both output VSIs are depicted in **Figure 9**. These voltages are measured with respect to a midpoint of the DC link. Then it is possible to calculate the sum of the voltage vectors:

$$\underline{v}_{sum} = \underline{v}_{A1} + \underline{v}_{B1} + \underline{v}_{C1} + \underline{v}_{A2} + \underline{v}_{B2} + \underline{v}_{C2} \tag{6}$$

$$\underline{v}_{sum} = v_{A1}e^{j0} + v_{B1}e^{-j\frac{2\pi}{3}} + v_{C1}e^{j\frac{2\pi}{3}} + v_{A2}e^{j\pi} + v_{B2}e^{j\frac{\pi}{3}} + v_{C2}e^{-j\frac{\pi}{3}} \tag{7}$$

Using the Euler's formula, $e^{j\alpha} = \cos\alpha + j\sin\alpha$, it is obtained as follows:

$$\underline{v}_{sum} = \left[v_{A1} - v_{A2} - \frac{1}{2}(v_{B1} - v_{B2}) - \frac{1}{2}(v_{C1} - v_{C2})\right] + j\frac{\sqrt{3}}{2}[-v_{B1} + v_{B2} + v_{C1} - v_{C2}] \tag{8}$$

that can be rewritten as follows:

$$\underline{v}_{sum} = \left(v_{A1A2} - \frac{1}{2}v_{B1B2} - \frac{1}{2}v_{C1C2}\right) + j\frac{\sqrt{3}}{2}(v_{C1C2} - v_{B1B2}) \tag{9}$$

Finally, the output voltage space vector can be defined as:

$$\underline{v}_o = \frac{2}{3}\underline{v}_{sum} = \frac{2}{3}\left(v_{A1A2} + v_{B1B2}e^{-j\frac{2\pi}{3}} + v_{C1C2}e^{j\frac{2\pi}{3}}\right) = v_o e^{j\theta} \tag{10}$$

where v_o is the magnitude and θ the angle of the space vector. The coefficient 2/3 is a scaling factor that has been added to keep constant magnitude of the vectors during the transformation [1].

In general, each inverter can produce eight independent voltage space vectors. Thus, there are a total of 64 vector combinations for the dual-inverter system, resulting in a space vector locus similar to a three-level neutral point clamped (NPC) inverter [4]. The space vectors for both inverters are shown in **Table 2**. A representation of the individual inverters space vectors is shown in **Figure 10**.

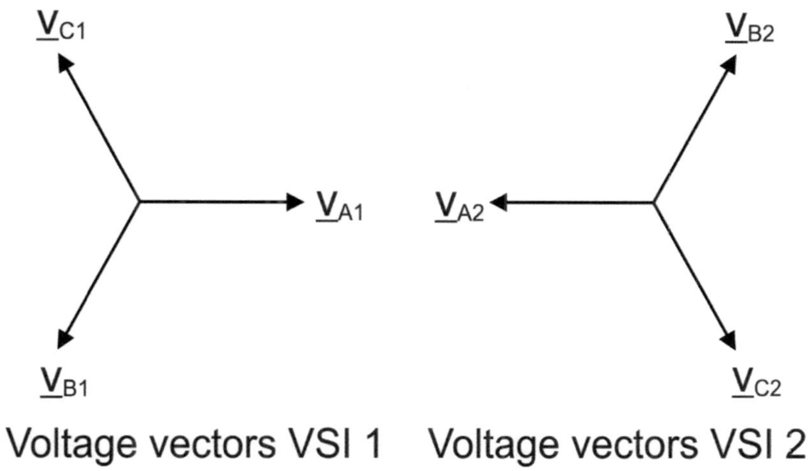

Figure 9. Voltages vectors of the individual inverters.

States of inverter 1 $[S_{A1}S_{B1}S_{C1}]$			
$V_1^1 = [100]$	$V_2^1 = [110]$	$V_3^1 = [010]$	$V_4^1 = [011]$
$V_5^1 = [001]$	$V_6^1 = [101]$	$V_7^1 = [111]$	$V_8^1 = [000]$
States of inverter 2 $[S_{A2}S_{B2}S_{C2}]$			
$V_1^2 = [100]$	$V_2^2 = [110]$	$V_3^2 = [010]$	$V_4^2 = [011]$
$V_5^2 = [001]$	$V_6^2 = [101]$	$V_7^2 = [111]$	$V_8^2 = [000]$

Table 2. Switching states of the individual inverters.

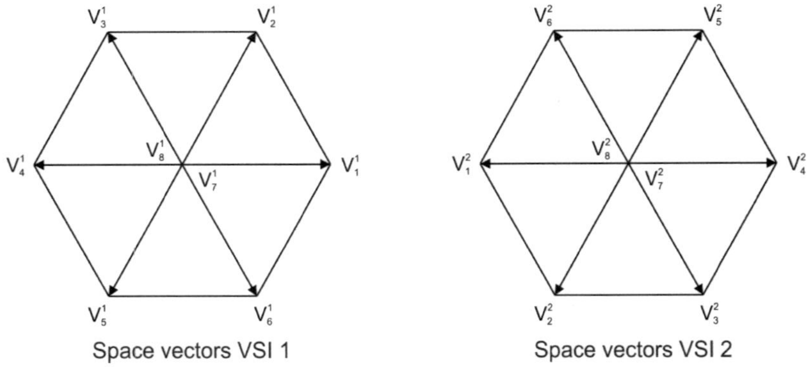

Figure 10. Space vectors representation of the individual inverters.

Let $V_{ij} = \left[V_i^1 V_j^2\right]$ with $i,\ j = 1\ldots8$, be the phase voltage vector combination of the dual-inverter system; hence, a diagram of the vector locations is shown in **Figure 11** [4], where the availability of redundant switching states for some voltage space vectors of the dual-inverter can be appreciated. This diagram is obtained by carrying out the vector sum of all the possible space vector combinations of inverters 1 and 2 (**Figure 10**).

The magnitude of the active space voltage vectors can be calculated considering that each phase load can be supplied with a voltage of $-V_{DC}$, 0 or $+V_{DC}$. For instance, the vector V_{14} applies $+V_{DC}$ to the phase-*a* load and $-V_{DC}$ to phase-*b* and phase-*c* loads. This results in:

$$\underline{V}_{14} = V_{DC} - V_{DC}e^{-j\frac{2\pi}{3}} - V_{DC}e^{j\frac{2\pi}{3}} = 2V_{DC} < 0^\circ[V] \tag{11}$$

The same procedure can be used to calculate the magnitude of each active space voltage vector of the dual-inverter system. This is summarized in **Table 3**.

As can be noted from **Table 3** and **Figure 11**, the six largest vectors have a magnitude of twice the DC-link voltage and have no redundancy. On the other hand, the higher redundancy is present for the lowest vectors each having six switching states available to produce the same output voltage.

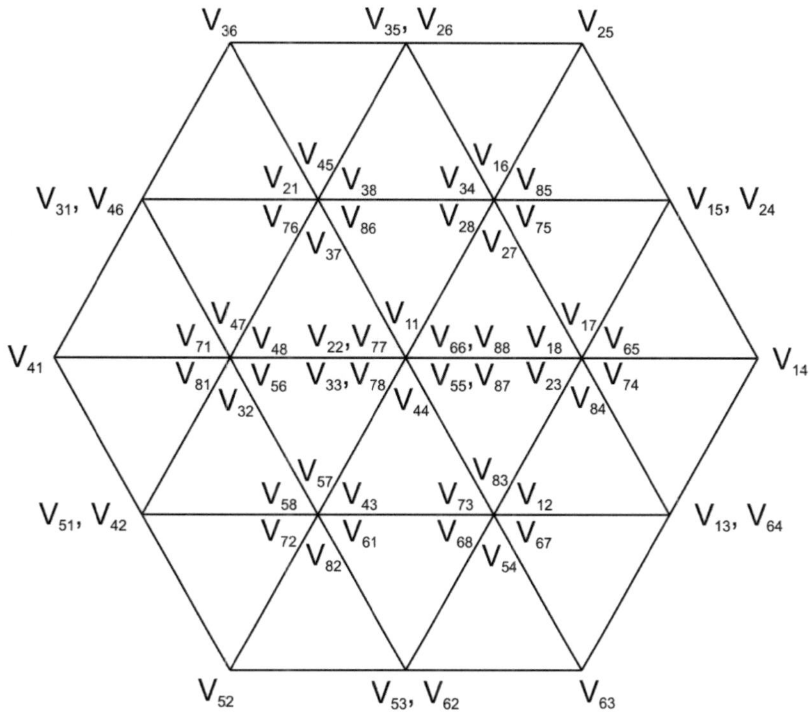

Figure 11. Space vector locations of the dual-inverter scheme.

	Space Vectors	Magnitude
Largest	$V_{14}, V_{25}, V_{36}, V_{41}, V_{52}, V_{63}$	$2V_{DC}$
Medium	$V_{15}, V_{24}, V_{35}, V_{26}, V_{31}, V_{46}, V_{51}, V_{42}, V_{53}, V_{62}, V_{13}, V_{64}$	$\sqrt{3}V_{DC}$
Lowest	$V_{65}, V_{17}, V_{18}, V_{23}, V_{84}, V_{74}, V_{85}, V_{16}, V_{34}, V_{28}, V_{27}, V_{75}$	V_{DC}
	$V_{38}, V_{45}, V_{21}, V_{76}, V_{37}, V_{86}, V_{48}, V_{47}, V_{71}, V_{81}, V_{32}, V_{56}$	
	$V_{43}, V_{57}, V_{58}, V_{72}, V_{82}, V_{61}, V_{12}, V_{83}, V_{73}, V_{68}, V_{54}, V_{67}$	

Table 3. Magnitude of the dual-inverter active space vectors.

3. Common-mode and zero sequence voltages in dual-inverters

As aforementioned, one of the main features of a dual-inverter circuit is that it allows reducing the common-mode voltage in the load then reducing the problems associated to it. On the other hand, it has been also stated that when supplying the dual-inverter with a single DC source, zero sequence current can flow in the load because of the generation of zero sequence voltage. In this section, a detailed explanation about the common-mode voltage and zero sequence voltage issues is given and solutions to reduce them are indicated.

3.1. Common-mode voltage

A typical three-phase sinusoidal power supply is balanced and symmetrical under normal conditions; that is, the sum of the three instantaneous voltages is zero. Thus, when supplying a balanced three-phase load, the voltage between an equivalent neutral point of the load and the neutral point of the voltage source is zero. Usually, the neutral point of the power source is grounded.

On the other hand, a three-phase PWM inverter is a source of asymmetrical voltages that switches a DC bus voltage (V_{DC}) into the three-phase terminals of the load, with a switching pattern that generates the proper fundamental frequency output voltage [2]. Since the output pole voltage of a two-level inverter, with respect to the negative rail of the DC bus, can be either $+V_{DC}$ or zero, it is not possible to have the three terminal voltages added to zero at any instant of time. The average voltage applied to the motor (over a cycle) is kept zero, but the instantaneous sum of the voltages at the load terminals is non-zero. Then, a voltage will appear between an equivalent neutral point of the load and the electrical ground of the system. This voltage is called common-mode voltage [30].

In an open-end load, such as depicted in **Figure 8**, the common-mode voltage is given by [23]:

$$v_{cm} = \frac{1}{6}(v_{A1G} + v_{B1G} + v_{C1G} + v_{A2G} + v_{B2G} + v_{C2G}) \tag{12}$$

where v_{AiG}, v_{BiG}, v_{CiG}, with $i = 1,\ 2$, are the pole voltages of each inverter with respect to the grounded neutral point of the power source (assuming in **Figure 8** that the DC voltage is provided by the rectification of a grounded AC system).

Because of the typically high switching frequency of a PWM inverter in the kHz range, the common-mode voltage has a high rate of change with respect to time (high dV/dt) and will generate common-mode currents due to capacitive couplings ($I_{cm} = C\,dV/dt$). Moreover, higher inverter switching frequencies will originate higher common-mode currents.

In AC motor drives, the capacitive couplings between different parts of a machine originate many potential paths for these common-mode currents to flow. The most common paths are [31] stator to rotor, stator winding to frame, rotor to shaft and shaft to frame. Therefore, the circulation of common-mode currents via the motor bearings back to the grounded stator case is possible. The so-called bearing currents have been found to be a major cause of premature bearing failure in PWM inverter motor drives [31]. Thus, a common-mode voltage reduction has been a topic of interest for many years.

3.1.1. Common-mode voltage in open-ended loads

One of the main advantages of an open-ended load connection is that it allows the possibility of reducing the common-mode voltage, then reducing the problems associated to it.

In the power converter topology shown in **Figure 8**, the pole voltages v_{AiG}, v_{BiG}, v_{CiG}, with $i = 1,\ 2$, can be expressed as follows:

$$\begin{aligned} v_{AiG} &= S_{Api}v_{pG} + S_{Ani}v_{nG} \\ v_{BiG} &= S_{Bpi}v_{pG} + S_{Bni}v_{nG} \\ v_{CiG} &= S_{Cpi}v_{pG} + S_{Cni}v_{nG} \end{aligned} \tag{13}$$

where v_{pG} and v_{nG} are the positive and negative rail voltages of the DC link with respect to the grounded neutral point of the power source, respectively. $S_{xpi}, S_{xni} \in \{0, 1\}$ with $x = A, B, C$, and $i = 1, 2$ are the switching functions of the inverter devices (0: switch closed, 1: switch opened) and $S_{xni} = 1 - S_{xpi}$ (due to the complementary operation of the upper and lower switches of each inverter leg). Hence, the common-mode voltage of Eq. (12) can be rewritten as follows:

$$\begin{aligned} v_{cm} = \frac{1}{6}\big[&(S_{Ap1} + S_{Bp1} + S_{Cp1} + S_{Ap2} + S_{Bp2} + S_{Cp2})v_{pG} \\ &+(S_{An1} + S_{Bn1} + S_{Cn1} + S_{An2} + S_{Bn2} + S_{Cn2})v_{nG}\big] \end{aligned} \tag{14}$$

Let $N_{sw} = S_{Ap1} + S_{Bp1} + S_{Cp1} + S_{Ap2} + S_{Bp2} + S_{Cp2}$, and considering that $v_{nG} = -v_{pG}$, thus

$$v_{cm} = \frac{1}{6}\left[N_{sw}v_{pG} + (6 - N_{sw})(-v_{pG})\right] = \frac{1}{6}v_{pG}[2N_{sw} - 6] \tag{15}$$

where N_{sw} is the number of upper inverter switches closed.

The squared RMS value of the common-mode voltage is as follows:

$$v^2_{cm_{RMS}} = \frac{1}{36T}\int_0^T [2N_{sw} - 6]^2 dt \tag{16}$$

where T is the period of v_{pG}. Further expansion yields:

$$36v^2_{cm_{RMS}} = (4N^2_{sw} - 24N_{sw} + 36)\frac{1}{T}\int_0^T dt = 4N^2_{sw} - 24N_{sw} + 36 \tag{17}$$

Differentiating Eq. (17) with respect to N_{sw} and equating to zero, it can be found that $v^2_{cm_{RMS}}$ (and implicitly $v_{cm_{RMS}}$) achieves a minimum value at $N_{sw} = 3$, which means that in order to reduce the RMS common-mode voltage at the machine terminals, only three upper inverter switches should be closed at each switching period.

This can be further investigated by considering a virtual midpoint of the DC link as a reference point. Then, Eq. (12) can be rewritten as follows:

$$v_{cm} = \frac{1}{6}(v_{A10} + v_{B10} + v_{C10} + v_{A20} + v_{B20} + v_{C20}) + v_{0G} = v_{cm0} + v_{0G} \tag{18}$$

where v_{cm0} is the common-mode voltage produced by the dual-inverter circuit with respect to a midpoint of the DC link and v_{0G} is the voltage between a midpoint of the DC source and the ground of the system.

The common-mode voltage produced by the 64 switching states combinations of the dual-inverter topology (v_{cm0}) can be calculated with Eq. (18) and is shown in **Table 4**.

Therefore, reducing the common-mode voltage in an open-end load is feasible if voltage vectors contained in the fourth row in **Table 4** are used.

However, it can be noted from **Table 3** and **Table 5** that the space vector combinations of the dual-inverter topology which eliminate the zero sequence voltage are not the same vectors which reduce the common-mode voltage.

V_{cm0}	**Voltage vector combinations**
$-V_{DC}/4$	V_{88}
$-V_{DC}/6$	$V_{85}, V_{83}, V_{81}, V_{58}, V_{38}, V_{18}$
$-V_{DC}/12$	$V_{84}, V_{86}, V_{82}, V_{55}, V_{35}, V_{33}, V_{51}, V_{31}$
	$V_{15}, V_{13}, V_{11}, V_{48}, V_{68}, V_{28}, V_{53}$
0	$V_{14}, V_{25}, V_{36}, V_{52}, V_{87}, V_{54}, V_{34}, V_{56}, V_{32}, V_{16}$
	$V_{12}, V_{45}, V_{43}, V_{41}, V_{65}, V_{63}, V_{23}, V_{21}, V_{78}, V_{61}$
$+V_{DC}/12$	$V_{17}, V_{57}, V_{37}, V_{44}, V_{46}, V_{64}, V_{24}, V_{42}$
	$V_{62}, V_{26}, V_{22}, V_{75}, V_{73}, V_{66}, V_{71}$
$+V_{DC}/6$	$V_{47}, V_{74}, V_{76}, V_{67}, V_{72}, V_{27}$
$+V_{DC}/4$	V_{77}

Table 4. Active space vectors producing null common-mode voltage.

V_{zs}	**Voltage vector combinations**
$-V_{DC}/2$	V_{87}
$-V_{DC}/3$	$V_{84}, V_{86}, V_{82}, V_{57}, V_{37}, V_{17}$
$-V_{DC}/6$	$V_{85}, V_{83}, V_{54}, V_{34}, V_{81}, V_{56}, V_{52}, V_{36}$
	$V_{32}, V_{47}, V_{14}, V_{16}, V_{12}, V_{67}, V_{27}$
0	$V_{88}, V_{55}, V_{53}, V_{35}, V_{33}, V_{44}, V_{51}, V_{31}, V_{46}, V_{42}$
	$V_{15}, V_{13}, V_{64}, V_{24}, V_{11}, V_{66}, V_{62}, V_{26}, V_{22}, V_{77}$
$+V_{DC}/6$	$V_{58}, V_{38}, V_{45}, V_{43}, V_{18}, V_{65}, V_{25}, V_{63}$
	$V_{23}, V_{74}, V_{41}, V_{61}, V_{21}, V_{76}, V_{72}$
$+V_{DC}/3$	$V_{48}, V_{68}, V_{82}, V_{75}, V_{73}, V_{71}$
$+V_{DC}/2$	V_{78}

Table 5. Zero sequence voltage contributions from different space vector combinations.

http://dx.doi.org/10.5772/intechopen.68450

3.2. Zero sequence voltage

It is well known that unbalanced three-phase voltages (or currents) can be transformed into three sets of voltage components [32]. These so-called symmetrical components are known as positive, negative and zero sequence components and can be schematically represented as shown in **Figure 12a**. Positive and negative sequence components correspond to three-phase balanced rotating phasors and zero sequence components are phasors with zero-phase shift angle. **Figure 12b** shows a decomposition of an unbalanced three-phase voltage into symmetrical voltage components.

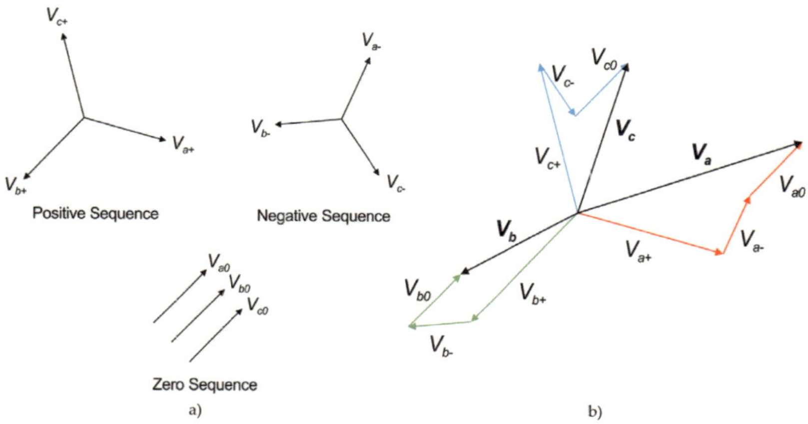

Figure 12. (a) Symmetrical components and (b) decomposition of unbalanced three-phase voltage into symmetrical components.

Unlike the positive and negative sequence currents, the main issue of the zero sequence currents is that they do not cancel but add up arithmetically at the neutral point of a four wire three-phase system, eventually overloading the neutral line or producing a higher neutral to ground voltage. Additionally, harmonic currents of any sequence circulating in an AC drive may give rise to increased RMS current, thus increasing the system losses, high current/voltage THD and machine over-heating and vibrations.

3.2.1. Zero sequence voltage in open-end loads

An open-end load supplied by a dual-inverter with a single DC source may suffer from zero sequence current caused by zero sequence voltage. This zero sequence voltage is produced because of the asymmetry of the instantaneous pulse width modulated phase voltages applied to the load phases (due to the voltage space vectors used). The zero sequence voltage is given by [18]:

$$v_{zs} = \frac{v_{A1A2} + v_{B1B2} + v_{C1C2}}{3} \tag{19}$$

or in terms of Eq. (5) as follows:

$$v_{zs} = \frac{1}{3} \sum_{k=a,\ b,\ c} v_{ph,ok} = \frac{v_{DC}}{3} \sum_{k=A,\ B,\ C} (S_{k1} - S_{k2}) \tag{20}$$

Thus, in order to make $v_{zs} = 0$, the following relationship must be satisfied:

$$\sum_{k=A,\ B,\ C} S_{k1} = \sum_{k=A,B,C} S_{k2} \tag{21}$$

Therefore, to eliminate the instantaneous zero sequence voltage in the load is necessary and sufficient to have the same number of upper (or lower) switches closed on both output inverters at every switching period.

By using Eq. (19), the zero sequence voltage contribution from the 64 space vector combinations of the dual-inverter topology can be calculated and is shown in **Table 5**. As can be noted, there are 20 space voltage vectors that do not produce zero sequence voltage, thus satisfying Eq. (21). Hence, in order to avoid the circulation of zero sequence current in the load, only these space voltage vector combinations could be used in the modulation strategy for the dual-inverter [24].

Moreover, from **Table 5** and **Figure 11**, it can be noted that there are two different but equivalent sets of active voltage vectors producing null zero sequence voltage (see **Table 6**), which could be used along with the zero voltage vectors: V_{11}, V_{22}, V_{33}, V_{44}, V_{55}, V_{66}, V_{77} and V_{88}.

Set 1	V_{15}	V_{35}	V_{31}	V_{51}	V_{53}	V_{13}
Set 2	V_{24}	V_{26}	V_{46}	V_{42}	V_{62}	V_{64}

Table 6. Active space vectors producing null zero sequence voltage.

Besides the use of space voltage vectors producing null v_{zs}, the occurrence of low order triplen harmonics in the load phase currents could be avoided performing a dynamic balance for the zero sequence current as proposed in Ref. [18]. This dynamic compensation method will be further discussed in Section 5.

4. Two two-level inverters fed by isolated DC sources

This circuit configuration is shown in **Figure 7**, where a standard two-level VSI is connected at each side of the load. The VSIs are supplied by isolated DC power sources. In general, the main characteristic of this topology is that circulation of zero sequence current in the load is avoided; however, it requires two isolation transformers to supply the DC sources increasing the cost and volume of system.

As the circuit configuration does not allow to have circulation of zero sequence current, the modulation strategy should aim to reduce only the common-mode voltage.

http://dx.doi.org/10.5772/intechopen.68450

4.1. SVM strategy for common-mode voltage reduction

It has been shown that an open-end load offers the possibility of reducing the common-mode voltage by using certain voltage space vector combinations of the dual-inverter [23], as shown in **Table 4**. The locus of the vectors that theoretically eliminate the common-mode voltage of the system is shown in **Figure 13**.

As can be noted in the locus, the vectors that reduce the common-mode voltage are the largest and some of the lowest, then depending on the output voltage requirement, the modulation for the dual-inverter could use the vectors of **Figure 13a** or **b**. Moreover, for the lowest vectors, there is switching states redundancy, opposite to the situation for the largest ones where all the voltage vectors can be produced by a unique switching state combination of the inverters.

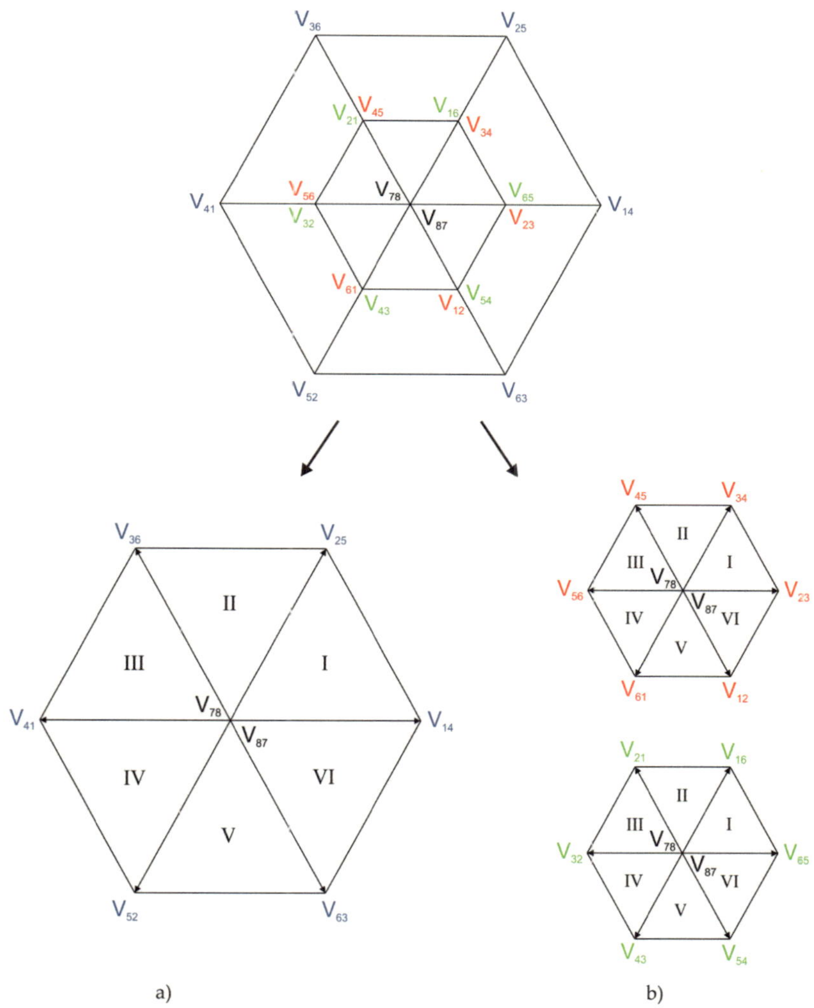

Figure 13. Locus of vectors for reduced common-mode voltage.

Symbol	Parameter	Value
R	Load resistance	4 Ω
L	Load inductance	6 mH
f_s	Switching frequency	10 kHz
f_o	Output voltage frequency	50 Hz
V_{DC}	DC voltage source	300 V

Table 7. Simulation parameters.

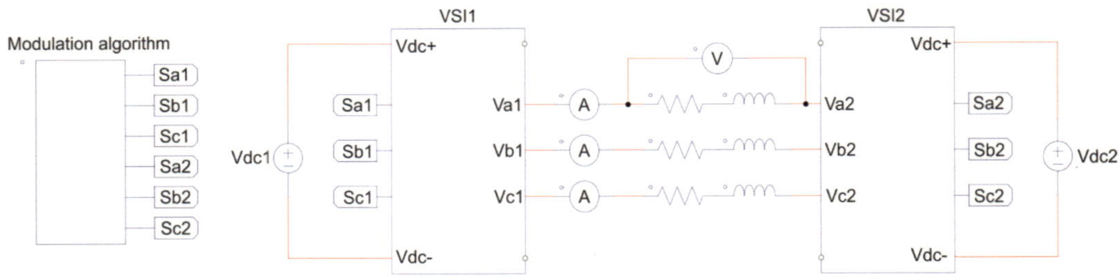

Figure 14. Circuit implemented in PSim.

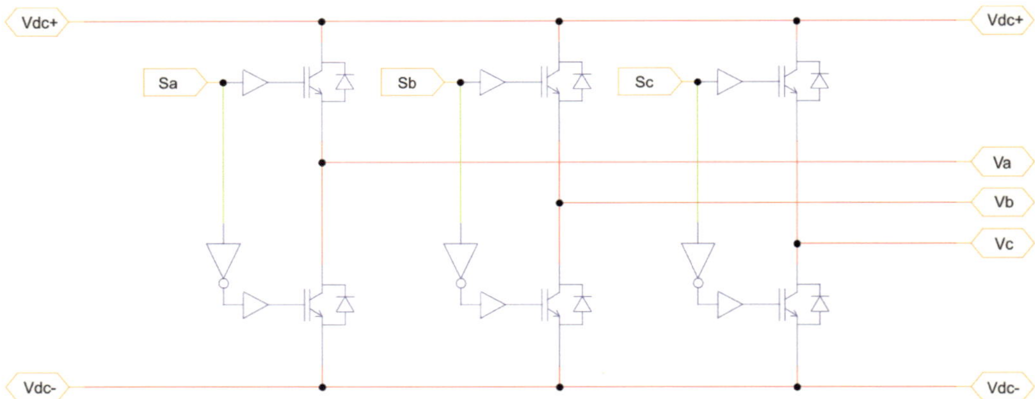

Figure 15. Individual inverter in PSim.

However, despite the aforementioned advantage of the lowest vectors, for the modulation strategy proposed, only the largest vectors will be used attending to maximize the output voltage and because the using of all the vectors available will complicate the modulation algorithm and the benefits in terms of current/voltage THD of applying the lowest space vectors are not significant. Once selected, the space vectors to be used, Eqs. (1) and (2), are valid for calculating the duty cycles and the switching sequence is the standard used in two-level VSIs (**Figure 5**).

4.1.1. Simulation results

The modulation strategy for common-mode voltage reduction has been simulated in PSim/Matlab simulation platform for the topology depicted in **Figure 7**, considering an R-L load and the parameters of **Table 7**. The modulation index used is the maximum possible without overmodulation. The circuit implemented in PSim is shown in **Figure 14** where the modulation algorithm is programmed in 'C' language in a special block provided by PSim software. The sub-circuits VSI1 and VSI2 (**Figure 14**) contain the standard two-level inverter shown in **Figure 15**. These circuits are used to simulate the required system and the obtained results (data tables) are then exported and plotted in Matlab environment.

The results are shown in **Figures 16–18**. **Figure 16** shows the output phase voltage (top) and its frequency spectrum (bottom). It can be noted that three levels are obtained in the load phase voltage. The frequency spectrum contains a fundamental component of 50 Hz and some low harmonic content around the switching frequency.

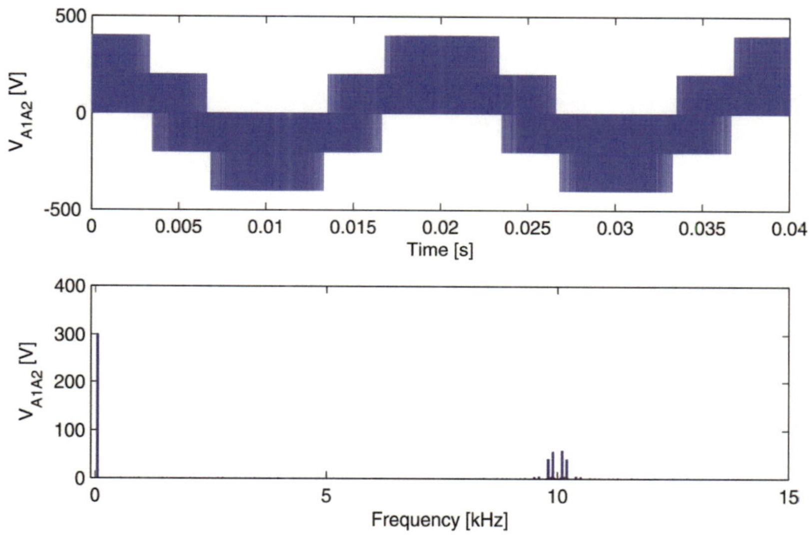

Figure 16. Output phase voltage (top) and its frequency spectrum (bottom) with SVM for reduced CMV and dual-inverter supplied with isolated DC sources.

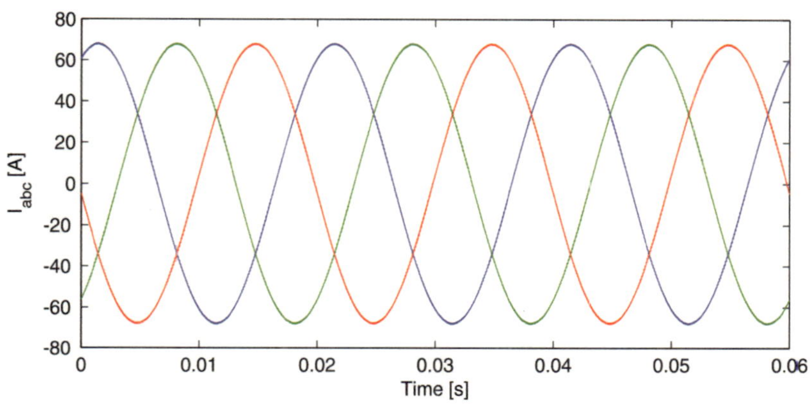

Figure 17. Output currents with SVM for reduced CMV and dual-inverter supplied with isolated DC sources.

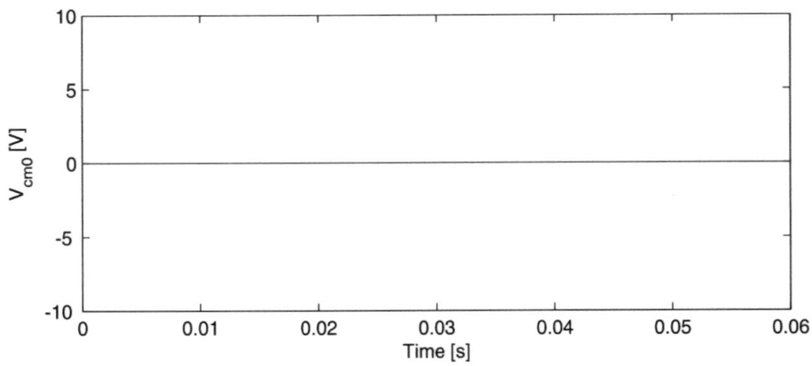

Figure 18. Common-mode voltage with SVM for reduced CMV and dual-inverter supplied with isolated DC sources.

The output currents are shown in **Figure 17** and the common-mode voltage is shown in **Figure 18**. It can be seen that the sinusoidal characteristic of the currents is due to the inductive nature of the load. Moreover, it can be noted that the CMV is eliminated due to the space vectors used in the modulation.

5. Two two-level inverters fed by a single DC source

The circuit configuration for a single DC source supplying a dual-inverter has been presented in **Figure 8**. The main disadvantage of this converter is that zero sequence current could circulate through the load due to the generation of output zero sequence voltage. Hence, a possible solution is to modulate the dual-inverter using only the space vectors that do not produce zero sequence voltage (**Table 5**). On the other hand, if the requirement is to reduce the CMV, the vectors of fourth row of **Table 4** can be used. However, it can be noted from **Table 4** and **Table 5** that the space vectors that reduce the CMV are not the same vectors that reduce the ZSV; therefore, to reduce both voltages at a time, a special modulation strategy is presented in this section.

5.1. SVM strategy for zero sequence voltage reduction

As mentioned above, the zero sequence voltage applied to the load can be eliminated by using certain voltage space vectors as shown in **Table 5**. Moreover, it has been mentioned that there are two equivalent sets of active vectors producing $v_{zs} = 0$ that can be used along with eight null vectors available in the dual-inverter. The locus of the vectors producing null v_{zs} is shown in **Figure 19**.

As can be seen in **Figure 19**, the hexagon is divided into six sectors and among the eight null vectors available, only six are finally used (three null vectors per set) [24]. Moreover, the null vectors should be mapped depending on the sector information [24] in order to reduce the commutations in a period. The mapping is shown in **Table 8**.

http://dx.doi.org/10.5772/intechopen.68450

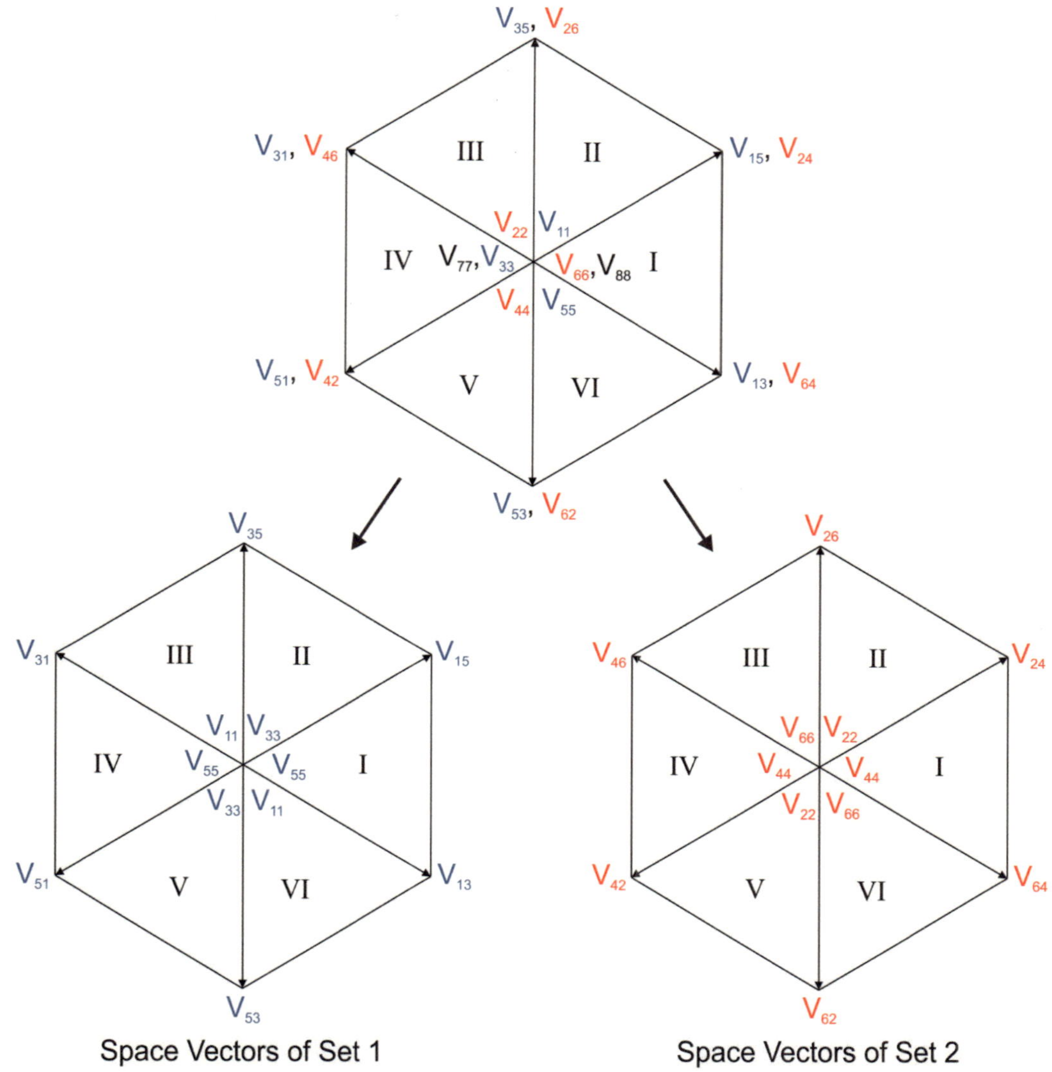

Figure 19. Locus of vectors producing null zero sequence voltage.

From **Tables 6** and **8**, it can be noted that in each sector, one of the inverters keeps clamped in a specific state and the other inverter commutates between three different switching states. This allows reducing the switching losses of the converter output stages.

The zero vectors V_{77} and V_{88} are not considered for this modulation scheme since none of the active vectors (**Table 6**) use the states $V_7 = [111]$ or $V_8 = [000]$ for the individual inverters. Hence, the application of V_{77} or V_{88} in the output stages will result in more commutations per period and thus in higher switching losses than the strategy proposed with the mapping of **Table 8**. However, these zero states are available if vectors redundancy is required.

Sector	I	II	III	IV	V	VI
Set 1 zero vectors	V_{55}	V_{33}	V_{11}	V_{55}	V_{33}	V_{11}
Set 2 zero vectors	V_{44}	V_{22}	V_{66}	V_{44}	V_{22}	V_{66}

Table 8. Mapping of zero vectors.

The space vector modulation presented allows reducing the output zero sequence voltage, then reducing the undesirable effects of the zero sequence currents. Moreover, there is voltage vectors redundancy thus allowing choosing between two equivalent sets of vectors producing the same phase voltage.

5.1.1. Simulation results

The modulation strategy for zero sequence voltage reduction has also been simulated in PSim considering the parameters of **Table 7** and the circuit shown in **Figure 14**, but considering a single DC source to supply both individual inverters. The modulation index used is the maximum possible without overmodulation. The results are shown in **Figures 20–22. Figure 20** shows the output phase voltage produced by the dual-inverter (top) and its frequency spectrum (bottom). The output PWM voltage obtained is unipolar and it has a fundamental voltage of 300 V, 50 Hz and harmonic content around the switching frequency (10 kHz). As can be seen, the frequency spectrum is similar to that of SVM for CMV reduction supplying the dual-inverter with a single DC source (**Figure 16**).

The output currents are shown in **Figure 21**. Due to the inductive nature of the load, the high frequency components of the voltage have a negligible effect on the current. **Figure 22** shows

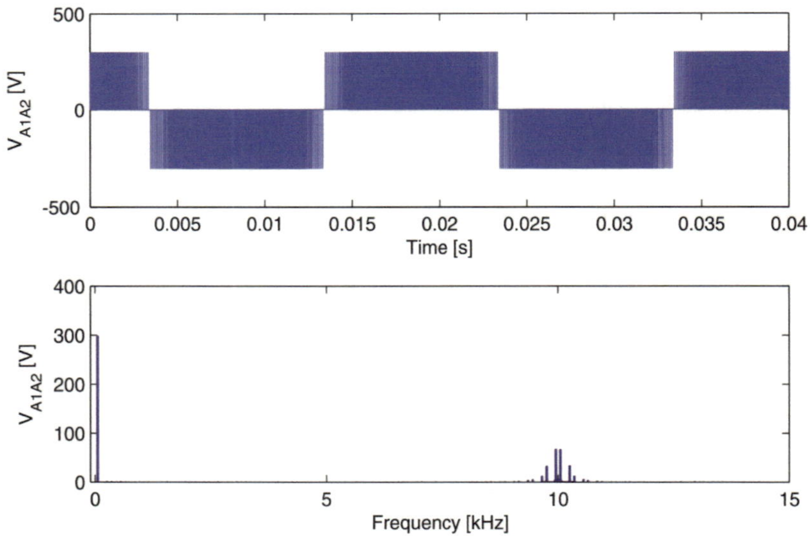

Figure 20. Output phase voltage (top) and its frequency spectrum (bottom) with SVM for reduced ZSV and dual-inverter supplied with single DC source.

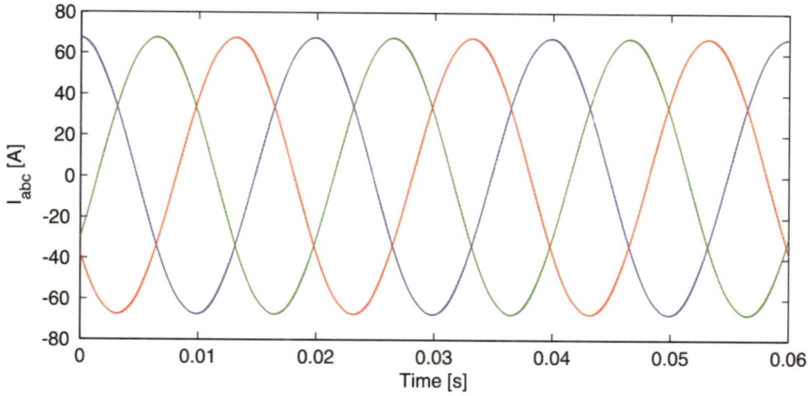

Figure 21. Output currents with SVM for reduced ZSV and dual-inverter supplied with single DC source.

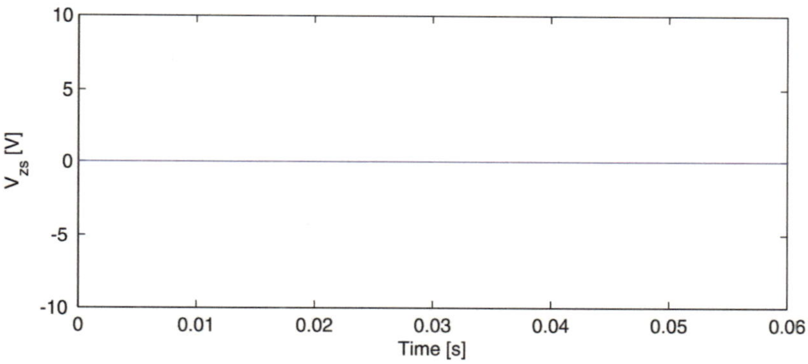

Figure 22. Zero sequence voltage with SVM for reduced ZSV and dual-inverter supplied with single DC source.

the zero sequence voltage that has been eliminated due to the space vectors used in the modulation of the dual-inverter.

5.2. SVM strategy for CMV reduction and ZSV compensation

It has been shown that an open-end load offers the possibility of reducing the common-mode voltage by using certain voltage space vector combinations of the dual-inverter [23], as shown in **Table 4**. Moreover, it has been mentioned that the vectors reducing the common-mode voltage will produce zero sequence voltage as can be noted in **Tables 4** and **5**. Therefore, compensation must be performed in order to avoid the circulation of zero sequence currents in the machine.

The compensation consists on eliminating the average zero sequence voltage within a sampling interval by forcing the zero sequence volt-seconds to zero [18]. This can be done by applying the null voltage vectors with unequal times [18], then modifying the standard switching pattern shown in **Figure 5** and commutating the inverters with the switching pattern of **Figure 23**.

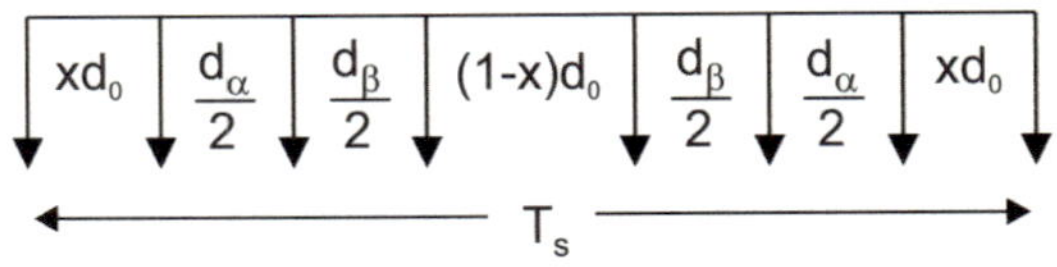

Figure 23. Modified switching sequence.

As it is known which space vector will be applied in every switching period, it can be known what the zero sequence voltage will be in every switching period as well. The value of x, which causes the cancellation of the zero sequence volt-seconds, is calculated at every sampling period to satisfy [18]:

$$2v_{zs1}xd_0 + v_{zs2}d_\alpha + v_{zs3}d_\beta + v_{zs4}(1-x)d_0 = 0 \tag{22}$$

where v_{zsk} with $k = 1, 2, 3, 4$, is the zero sequence voltage value (calculated with Eq. (10)) at intervals xd_0, d_α, d_β and $(1-x)d_0$, respectively.

The x coefficient must be calculated at every switching period to allow a correct reduction of the output zero sequence volt-seconds. The modulation strategy that is presented reduces the common-mode voltage produced by the output VSIs of the power converter and compensates the occurrence of zero sequence voltage.

5.2.1. Simulation results

The modulation for common-mode voltage reduction and zero sequence voltage compensation is simulated considering the parameters shown in **Table 7**. The modulation results in a bipolar PWM waveform can be seen in the output voltage of **Figure 24** (top). **Figure 24** (bottom) shows the frequency spectrum of the output voltage where the high frequency components (around

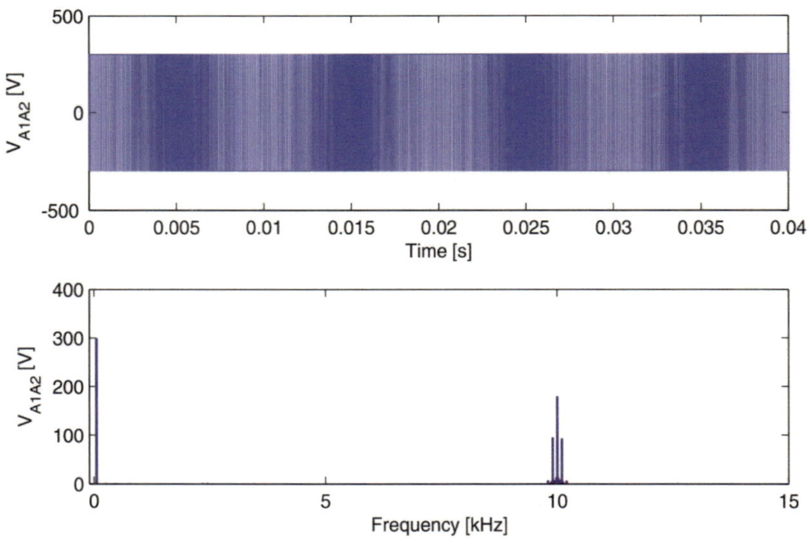

Figure 24. Output phase voltage (top) and its frequency spectrum (bottom) with SVM for CMV-ZSV reduction and dual-inverter supplied with single DC source.

10 kHz) are of higher magnitude than the modulation for zero sequence voltage reduction (**Figure 20**). This is due to the bipolarity of the PWM.

The output currents are shown in **Figure 25**, where it can be noted that zero sequence components are not present due to compensation method used. **Figure 26** shows the common-mode voltage that has been eliminated due to the space vectors used in the modulation.

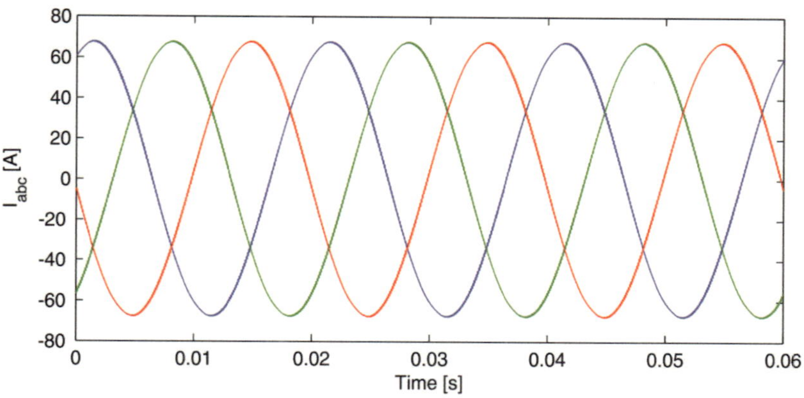

Figure 25. Output currents with SVM for CMV-ZSV reduction and dual-inverter supplied with single DC source.

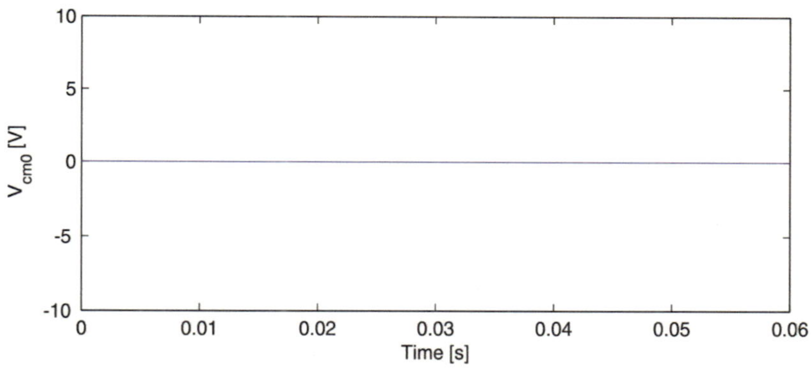

Figure 26. Common-mode voltage with SVM for CMV-ZSV reduction and dual-inverter supplied with single DC source.

6. Multilevel topologies

Several multilevel power converters have been developed for open-end loads, specifically open-end winding drive. For example, **Figure 27a** shows a three-level inverter [33] and **Figure 27b** shows a five-level inverter [34]. It can be noted that the five-level inverter presents the same topology of the three-level inverter but considering isolated DC supplies. The main advantage of the multilevel topologies is that the machine phase voltage presents lower voltage distortion increasing the performance of the drive but on the other hand, the complexity and cost of the system is also increased.

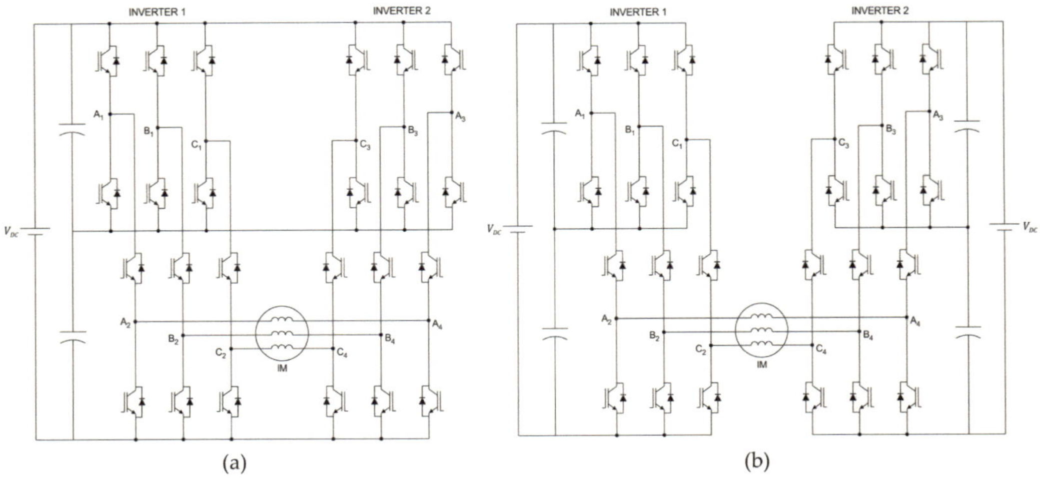

Figure 27. (a) Three-level inverter and (b) five-level inverter for open-end winding AC machine drives.

6.1. Carrier-based modulation strategy

In a standard two-level inverter, a sinusoidal (carrier-based) pulse width modulation (SPWM) strategy requires the comparison of a triangular wave (carrier) with a sinusoidal reference signal (**Figure 3**). However, in multilevel inverters, more than one carrier signals are needed to perform a SPWM strategy [35]. Considering the five-level inverter of **Figure 27b**, four triangular carriers are required to be compared with three sinusoidal reference signals (one reference signal for each phase). The carriers and a reference signal for one phase are shown in **Figure 28**. The reference

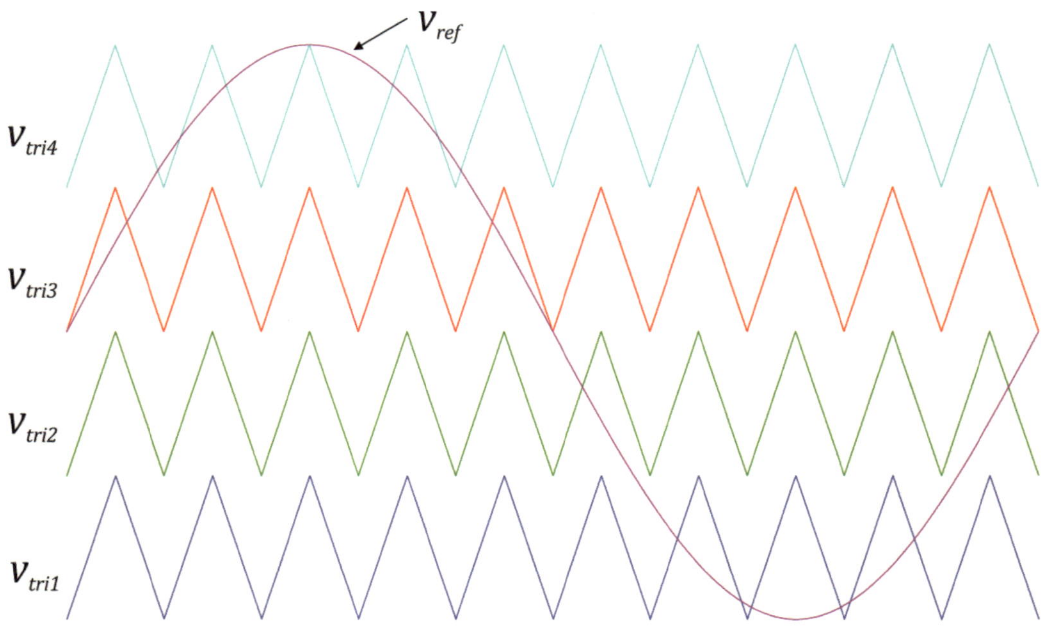

Figure 28. Signals used in a five-level carrier-based modulation strategy (one phase of the inverter).

http://dx.doi.org/10.5772/intechopen.68450

signals of the other two phases are phase-shifted ±120° and the control logic for triggering the power devices of the converter is similar to that shown in Section 1.

6.1.1. Simulation results

A SPWM strategy for a five-level dual-inverter is simulated in PSim platform. The circuit implemented in PSim is basically the same as **Figure 14**, but the sub-circuits inside VSI1 and VSI2 are those corresponding to the five-level inverter shown in **Figure 27b**. The simulation considers the parameters of **Table 7**.

Figure 29 shows the output phase voltage (top) and its frequency spectrum (bottom). The benefit of five-level operation in terms of voltage quality can be noted in the almost negligible harmonic content of the waveform. The output currents are shown in **Figure 30** which presents a sinusoidal waveform due to the inductive nature of the load.

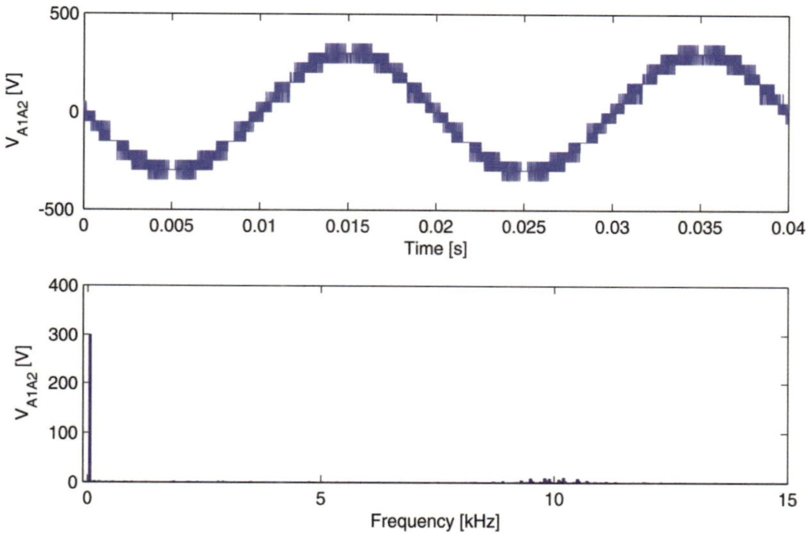

Figure 29. Output phase voltage (top) and its frequency spectrum (bottom) with SPWM for five-level inverter.

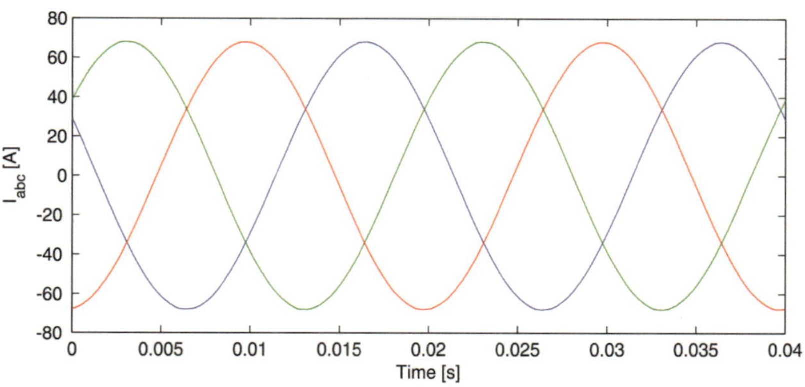

Figure 30. Output currents with SPWM for five-level inverter.

7. Conclusion

The dual-inverter circuit to supply open-ended loads has been presented. The possibility of supplying the VSI either from different or the same DC voltage sources have been emphasized, stating the main advantages and disadvantages of both alternatives. The main features of the topology have been studied and different modulation strategies have been developed attending the reduction of common-mode voltage when the topology uses different DC power sources and the reduction of common-mode voltage and/or zero sequence voltage when the topology uses a single DC power supply for both VSIs. Simulation results showing the performance of the modulation strategies proposed have been presented where isolated and non-isolated DC power supplies have been considered. Moreover, multilevel dual-inverter circuits have been discussed as an alternative to produce higher quality voltages and a standard SPWM strategy has been commented and simulated. The results are encouraging and demonstrate the applicability of the topology for open-end terminal loads.

Acknowledgements

This work was funded by The Chilean Research Fondecyt Grant 1151325 and by CONICYT/FONDAP/15110019. The financial support given by University of Bío-Bío Research Project N°GI160510 EF is also acknowledged.

Nomenclature

AC	Alternating-current
CMV	Common-mode voltage
DC	Direct-current
EMF	Electromotive force
IGBT	Insulated gate bipolar transistor
NPC	Neutral point clamped
PMSM	Permanent magnet synchronous motor
PWM	Pulse width modulation
SPWM	Sinusoidal (carrier-based) pulse width modulation
SVM	Space vector modulation
THD	Total harmonic distortion
VSI	Voltage source inverter
ZSV	Zero sequence voltage

http://dx.doi.org/10.5772/intechopen.68450

Author details

Javier Riedemann Aros[1]*, Rubén Peña Guíñez[2] and Ramón Blasco Gimenez[3]

*Address all correspondence to: jriedema@ubiobio.cl

1 Department of Electrical and Electronic Engineering, University of Bío-Bío, Concepción, Chile

2 Department of Electrical Engineering, University of Concepción, Concepción, Chile

3 Department of System Engineering and Control, Universitat Politecnica de Valencia, Valencia, Spain

References

[1] Rashid M. Power Electronics: Circuits, Devices, and Applications. 3rd ed. 2003. ISBN: 0131011405

[2] Mohan N, Undeland T, Robbins W. Power Electronics: Converters, Applications, and Design. 3rd ed. 2002. ISBN: 0471226939

[3] Mohapatra KK, Gopakumar K, Somasekhar VT, Umanand L. A harmonic elimination and suppression scheme for an open-end winding induction motor drive. IEEE Transactions on Industrial Electronics. December 2003;**50**(6): 1187-1198

[4] George DS, Baiju MR. Decoupled random modulation technique for an open-end winding induction motor based 3-level inverter. IEEE Symposium on Industrial Electronics & Applications. 4–6 October 2009;**2**: 1022-1027

[5] Somasekhar VT, Gopakumar K, Shivakumar EG, Sinha SK. A space vector modulation scheme for a dual two level inverter fed open end winding induction motor drive for the elimination of zero sequence currents. European Power Electronic and Drives Journal. March 2002;**12**(2)

[6] Oleschuk V, Sizov A, Bose BK, Stankovic AM. Phase-Shift-Based Synchronous Modulation of Dual Inverters for an Open-End Winding Motor Drive with Elimination of Zero Sequence Currents. International Conference on Power Electronics and Drives Systems. 16–18 January 2006;**1**: 325-330

[7] Wang Y, Panda D, Lipo TA, Di Pan. Open-Winding Power Conversion Systems Fed by Half-Controlled Converters. IEEE Transactions on Power Electronics. May 2013;**28**(5):2427-2436

[8] Wang Y, Lipo TA, Pan D. Robust operation of double-output AC machine drive. 2011 IEEE 8th International Conference on Power Electronics and ECCE Asia (ICPE & ECCE); 30 May–3 June 2011; pp.140-144

[9] Gerrits T, Wijnands CGE, Paulides JJH, Duarte JL, Lomonova EA. Operating range advancement of a dual-inverter driven machine through voltage range enhancement. IEEE Transactions on Transportation Electrification. September 2016; **2**(3): 322-334

[10] Kiadehi AD, Drissi KEK, Pasquier C. Voltage THD Reduction for Dual-Inverter Fed Open-End Load With Isolated DC Sources. IEEE Transactions on Industrial Electronics. March 2017; **64**(3): 2102-2111

[11] Chen M, Sun DA. Unified space vector pulse width modulation for dual two-level inverter system. IEEE Transactions on Power Electronics. February 2017; **32**(2): 889-893

[12] Ramachandrasekhar K, Mohan S, Srinivas S. An improved PWM for a dual two-level inverter fed open-end winding induction motor drive. In: 2010 XIX International Conference on Electrical Machines (ICEM); 6–8 September 2010; pp.1-6

[13] Zhu B, Prasanna UR, Rajashekara K, Kubo H. Comparative study of PWM strategies for three-phase open-end winding induction motor drives," In: 2014 International Power Electronics Conference (IPEC-Hiroshima 2014–ECCE-ASIA); 18–21 May 2014; pp. 395-402

[14] Bodo N, Levi E, Jones M. Investigation of carrier-based PWM techniques for a five-phase open-end winding drive topology. IEEE Transactions on Industrial Electronics. May 2013; **60**(5): 2054-2065

[15] Bodo N, Levi E, Jones M. "Enhanced decoupling PWM methods for five-phase open-end winding drives. In: 20th International Conference on Electrical Machines (ICEM '12);. 2–5 September 2012; pp. 939-945

[16] Patkar F, Levi E, Jones M. A six-phase multilevel space vector PWM algorithm for a dual-inverter supplied drive 6th IET International Conference on Power Electronics, Machines and Drives (PEMD 2012); 27–29 March 2012; pp.1-5

[17] Darijevic M, Jones M, Dordevic O, Levi E. Decoupled PWM control of a dual-inverter four-level five-phase drive. IEEE Transactions on Power Electronics. May 2017; **32**(5): 3719-3730

[18] Somasekhar VT, Srinivas S, Prakash Reddy B, Nagarjuna Reddy C, Sivakumar K. Pulse width-modulated switching strategy for the dynamic balancing of zero-sequence current for a dual-inverter fed open-end winding induction motor drive. IET Electric Power Applications. July 2007; **1**(4): 591-600

[19] Somasekhar VT, Srinivas S, Kumar KK. Effect of zero-vector placement in a dual-inverter fed open-end winding induction-motor drive with a decoupled space-vector PWM strategy. IEEE Transactions on Industrial Electronics. June 2008; **55**(6): 2497-2505

[20] Senicar F, Junge C, Gruber S, Soter S. Zero sequence current elimination for dual-inverter fed machines with open-end windings." In: 36th Annual Conference on IEEE Industrial Electronics Society (IECON '10); 7–10 November 2010; pp. 853-856

[21] An Q, Liu J, Peng Z, Sun L, Sun L. Dual-Space Vector Control of Open-End Winding Permanent Magnet Synchronous Motor Drive Fed by Dual Inverter. IEEE Transactions on Power Electronics. December 2016; **31**(12): 8329-8342

[22] Said NAM, Priestley M, Dutta R., Fletcher JE. Torque ripple minimization in dual inverter open-end winding PMSM drives with non-sinusoidal back-EMFs by harmonic current suppression. 42nd Annual Conference of the IEEE Industrial Electronics Society 23-26 October 2016 (IECON 2016); Florence, IEEE, 2016. pp. 2975-2980

[23] Srinivasan P, Reddy BV, Somasekhar VT. PWM switching strategy for the elimination of common mode voltage of a two-level inverter drive with an open-end winding induction motor configuration. In: Joint International Conference on Power Electronics, Drives and Energy Systems (PEDES) & 2010 Power India, 2010; 20–23 December 2010; pp.1-6

[24] Somani A, Gupta RK, Mohapatra KK, Basu K, Mohan N. Modulation strategies for direct-link drive for open-end winding AC machines. IEEE International Electric Machines and Drives Conference, 2009. (IEMDC '09); 3–6 May 2009; pp. 1863-1868

[25] Somani A, Gupta RK, Mohapatra KK, Mohan N. A minimum-switch direct-link drive with common-mode voltage suppression and active filtering for open-end winding AC machines. In: IEEE Power Electronics Specialists Conference, 2008. (PESC '08); 15–19 June 2008; pp. 2889-2893

[26] Bodo N, Jones M, Levi E. PWM techniques for an open-end winding fivephase drive with a single DC source supply. In: 38th Annual Conference on IEEE Industrial Electronics Society (IECON 2012); 25–28 October 2012; pp. 3641-3646

[27] Bodo N, Jones M, Levi E, Iqbal A. PWM techniques for an open-end winding seven-phase drive with a single DC source supply. In: 39th Annual Conference of the IEEE Industrial Electronics Society (IECON '13); 10–13 November 2013; pp. 5209-5214

[28] Rajeevan PP, Abu-Rub H, Iqbal A, Gopakumar K. Common mode voltage elimination scheme for dual-inverter fed five phase AC drives with open-end stator windings. In: 2013 IEEE International Conference on Industrial Technology (ICIT); 25–28 February 2013; pp. 1680-1685

[29] Bodo N, Jones M, Levi E. A space vector PWM with common-mode voltage elimination for open-end winding five-phase drives with a single DC supply. IEEE Transactions on Industrial Electronics. May 2014; **61**(5): 2197-2207

[30] Hongfei M, Dianguo X, Lijie M. Suppression techniques of common-mode voltage generated by voltage source PWM inverter. The 4th International Power Electronics and Motion Control Conference, 2004. (IPEMC 2004); 14–16 August 2004; Vol. 3. pp. 1533-1538

[31] Shaotang Chen, Lipo, TA, Fitzgerald D. Modeling of motor bearing currents in PWM inverter drives. In: Conference Record of the 1995 IEEE Industry Applications Conference, 1995. Thirtieth IAS Annual Meeting (IAS '95); 8–12 October 1995; Vol.1. pp. 388-393

[32] Costa LLH, Serni PJA, Marafao FP. An analysis of Generalized Symmetrical Components in non sinusoidal three phase systems. In: 2011 Brazilian Power Electronics Conference (COBEP); 11–15 September 2011; pp. 502-507

[33] Kanchan RS, Tekwani PN, Gopakumar K. Three-level inverter scheme with common mode voltage elimination and DC-link capacitor voltage balancing for an open end winding induction motor drive. In: 2005 IEEE Electric Machines and Drives Conference 15 May 2005. IEEE

[34] Baiju MR, Gopakumar K, Mohapatra KK, Somasekhar VT, Umanand L. Five-level inverter voltage-space phasor generation for an open-end winding induction motor drive. In: IEE Proceedings Electric Power Applications; 9 September 2003;150(5);531-538

[35] Holmes DG, Lipo TA. Pulse Width Modulation for Power Converters: Principles and Practice. Hoboken, NJ: John Wiley; 2003

Chapter 6

Real-Time Implementation of the Advanced Control of the Three-Phase Induction Machine Based on Power Inverters

Marian Gaiceanu

Additional information is available at the end of the chapter

http://dx.doi.org/10.5772/intechopen.68217

Abstract

The fast growing development of both the numerical equipment and power electronics allows the rapid prototyping of the innovating idea. The objective of this chapter is to put into evidence the teaching aspects through the applicative research in the field of the electric drives. The chapter provides the basic and advanced aspects of the electric drives control based on the most used electrical machine: three-phase induction motor (IM). The research work is presented in didactical way, starting with the conventional vector control, followed by the integration of the model reference adaptive control into the specific IM-based drive. The verified numerical simulation results push the research process through the implementation way. In order to increase the IM drives efficiency, the real-time implementation of the most commonly used modulation techniques is provided. Based on the dSpace platform, interfaced by ControlDesk, the experimental results are obtained. Both the performances of the cascaded control and model reference adaptive control are shown.

Keywords: DS1104, Matlab®, Simulink, PWM, THPWM, OHPWM, power inverter, efficiency, sustainability

1. Introduction

The technical literature subject to teaching of the adjustable AC drives, at the undergraduate and post-graduate levels, offers various techniques of learning based on the computer tools [1–5]. The fast growing of the digital technology conducts to the inherent replacement of the analog control by the numerical ones. A successful AC drive-based teaching tool should include at least the parameter identification/estimation task, the adequate control, and one friendly interface between the student and the computer. One of the first approaches of including the computer as an interactive learning environment and real-time implementation in adjustable drive teaching process is presented in Refs. [1, 2]. The virtual education environment allows the hands-on exercises to be tested and solved in a real-time environment [1, 2].

By using the obtained values from the no-load and short-circuit tests and the Matlab/Simulink software facilities [3], the electrical and mechanical parameters of the three-phase induction motor are determined. In order to teach and well understand the complex electromechanical phenomena of an adjustable speed drive based on the IM, the authors [4] use the Matlab/ Simulink software as a teaching tool. The teaching tool based on the intelligent control is presented by the authors of the chapter [5]. The fuzzy logic is one of the controlled ways in order to avoid the parameter identification/estimation task of the AC drive. The modern electric drives include the increase of the efficiency in the control design process. Despite the above presented state of the art, the author of this chapter provides an original robust model reference adaptive control of the three-phase induction motor in which the parameters of the controller are provided on-line; the parameters being adapted through the on-line estimator.

A major key-enabled technology for sustainability of the electrical energy is the enhancement of the efficiency characteristics in power inverter applications. Therefore, both energy saving potential and optimization of the energy consumption should be explored [6]. For the electric drives area, combining variable speed drives with modulation techniques could be one way of sustainability of both the electrical energy producers and consumers. Another way of assuring the sustainability of the electric drives is the use of the optimal control theory [6]. The power electronics is other key enabling technology in energy efficiency, as well in production, distribution and energy transport [5]. This chapter provides solutions both to increase the power quality and efficiency of the static power inverter. The chapter offers, in the didactical manner, the basic theoretical concerns regarding the static conversion by means of the power inverters, mathematical modelling of the power inverter, the modulation strategies, the numerical simulation and the experimental results for the presented modulation techniques. The chapter is addressed to the future and current changeable actors: the students, researchers and engineers. The chapter contains the basic concepts and techniques to design, simulate and implement the efficient power inverters through a Matlab-Simulink well-structured technical guide. In this manner, the chapter is addressed both to the students and researchers in the field of the electric drives. The mathematical model of the power inverter combined with the rotor field vector control of the three-phase induction machine (IM) is provided in Section 2. In Section 3, an overview of the modulation techniques is given with the purpose to point out the development trend in the power inverters technology. Moreover, the original model reference adaptive control of the three-phase induction machine is provided in Section 4.

2. Field-oriented control of the three-phase induction machine

In industry, the most commonly used electric drive is based on the three-phase induction machine (IM). It is well known that the mathematical model of the three-phase induction

http://dx.doi.org/10.5772/intechopen.68217

machine is nonlinear (the nonlinearities type are product, saturation and hysteresis), being a multivariable coupling of the control structure. For the same size, weight and inertia, the performances of the IM are higher than that of the DC motor. Therefore, the efficiency and the maximum speed are superior to that of the DC motor, in the lower price. From the electrical drive point of view, there are mainly two types of control: scalar and vectorial. At the constant flux, due to the decoupling control, the performances of the vector-control IM drives are better than of the scalar drives.

The field-oriented control concept allows independent control of the mechanical and electrical circuits through the stator active and reactive components. The field-oriented control could be in direct form or indirect form, depending on the flux vector position determination. The invariance propriety of the electromagnetic torque to the reference frames conduct to the three basic field-oriented schemes: stator field-oriented control, rotor field-oriented control, air-gap or arbitrary field-oriented control. The most used one is the rotor-magnetizing current reference frame [6] that rotates synchronously with the angular speed of the rotor-magnetizing current phasor. In order to apply this type of the vector control, the phasor of the rotor-magnetizing current should be known. This requirement assumes the real-time calculus of the position and the modulus of the magnetizing current phasor.

Due to the orthogonality between the stator and rotor magnetic fields, by neglecting the saturation, the magnetic flux depends on the stator current, without being influenced by the rotor current. By aligning the rotor-magnetizing phasor with the *d* axis of the synchronously (*d*, *q*) quadrature reference frame, the components of it are as follows:

$$\lambda_{qr} = 0,\ \lambda_{dr} = \lambda_r = \lambda = ct. \tag{1}$$

In rotor field reference frame, the mathematical model of the three-phase induction machine is as follows [6, 7]:

$$\begin{aligned}
\sigma T_s \frac{di_{ds}}{dt} + i_{ds} &= \frac{v_{ds}}{R_s} - (1-\sigma) T_s \frac{di_{mR}}{dt} + \sigma T_s \omega_{mR} i_{qs} \\
\sigma T_s \frac{di_{qs}}{dt} + i_{qs} &= \frac{v_{qs}}{R_s} - (1-\sigma) T_s \omega_{mR} i_{mR} - \sigma T_s \omega_{mR} i_{ds} \\
T_r \frac{di_{mR}}{dt} + i_{mR} &= i_{ds} \\
J \frac{d\omega_r}{dt} &= k_T i_{mR} i_{qs} - T_L \\
\omega_{mR} = \omega_r + \omega_{sl},\ \omega_{sl} &= \frac{R_r}{L_m} \frac{i_{qs}}{i_{mR}}
\end{aligned} \tag{2}$$

By taking into account the above mentioned system equations (2), the coupling between the d and q voltage components could be noticed. By adding the adequate feedforward electromotive voltage components, E_d and E_q, the IM mathematical model could be decoupled (**Figure 1**) [6, 7]:

$$v_{ds}^* = K_i \int_0^t e_d dt + K_p e_d - \omega_{mR} \sigma L_s i_{ds} - E_d$$
$$v_{qs}^* = K_i \int_0^t e_q dt + K_p e_q + \omega_{mR} \sigma L_s i_{ds} + E_q \tag{3}$$

$$E_q = \omega_{mR} \frac{L_m}{L_r} \lambda_{dr}, \; E_d = \omega_{mR} \frac{L_m}{L_r} \lambda_{qr} = 0, \tag{4}$$

In **Figure 2**, the block diagram of the Field Oriented Control (FOC) based on three-phase induction motor in rotor field magnetizing current reference frame is depicted:

The mathematical model of the rotor-magnetizing field of the three-phase induction machine consists of the electrical and mechanical transfer functions (**Figure 1**).

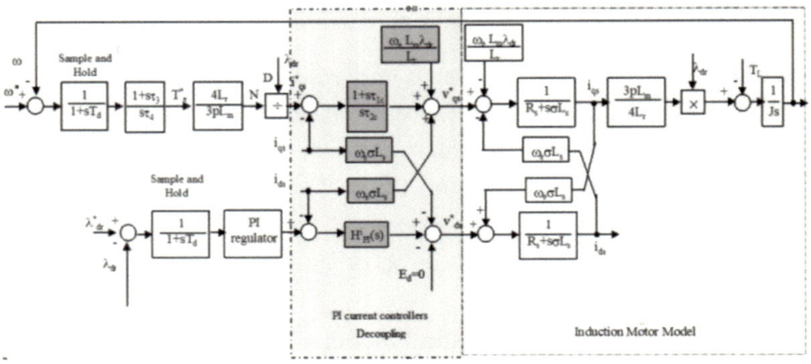

Figure 1. FOC of IM in rotor field-magnetizing current reference frame.

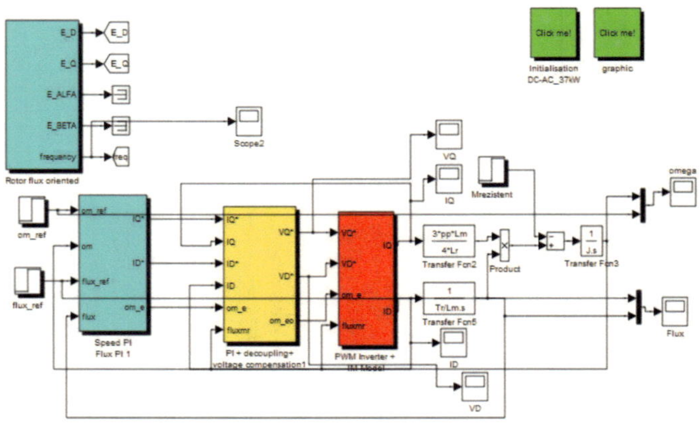

Figure 2. Simulink block diagram of the rotor field vector control of the three-phase induction machine.

In order to supply adequate power to the IM, the three-phase power inverter is necessary. By comparing the v^*_{ds} (v^*_{qs}) *d*-axis stator reference voltage and *q*-axis stator reference voltage with the carrier signal, the adequate switching states are delivered. This task is accomplished by the modulators.

In **Figure 2**, the Simulink implementation of the rotor field vector control of the three-phase induction machine is shown.

The Proportional-Integral (PI) speed and flux controllers are shown in **Figure 3**.

In **Figure 4**, the rotor magnetic flux position is shown [8]. In the Simulink group shown in **Figure 5**, the PI *d-q* current controllers and voltage decoupling terms are added.

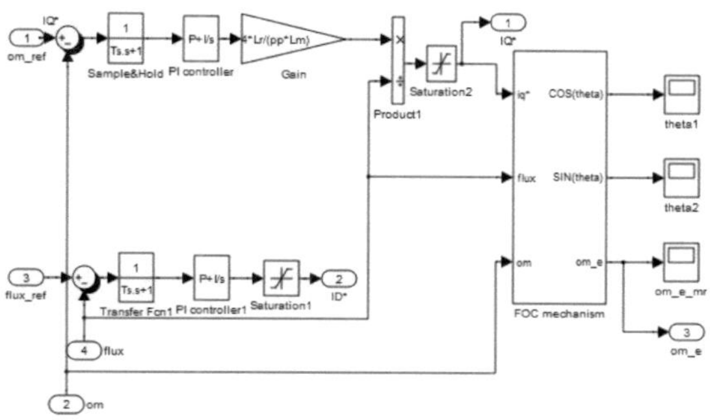

Figure 3. The control side of the FOC with IM.

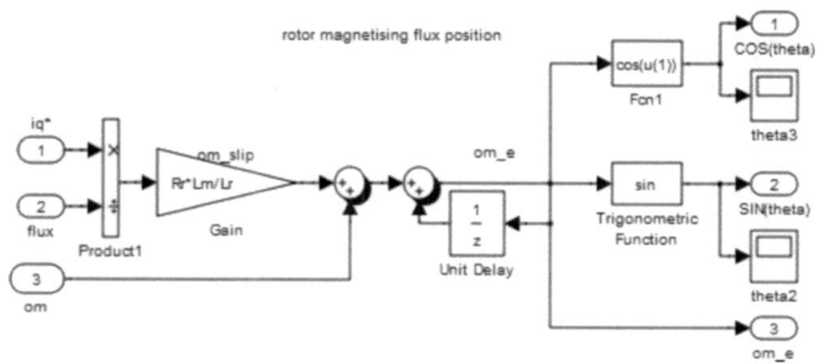

Figure 4. Determination of the synchronous reference frame position.

In **Figure 6**, both the inverter transfer function and the stator equivalent *d-q* windings are shown.

By using the Simulink program described in this chapter, the references and the output state variables are presented in **Figure 7**. The inner torque control loop is tuned by using the modulus criterion. The speed outer loop control is performed based on the symmetrical

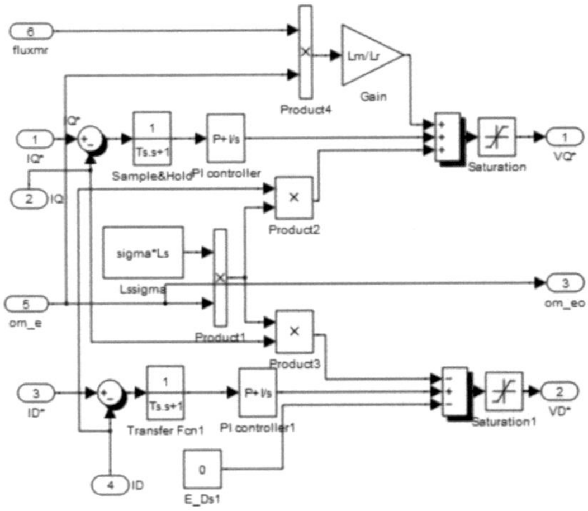

Figure 5. Current control and the voltage decoupling block.

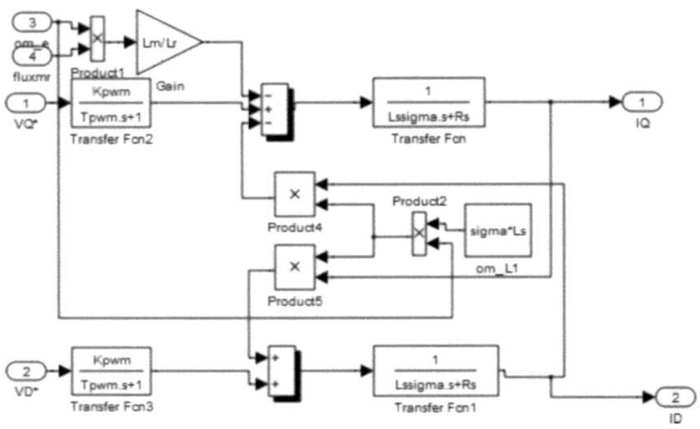

Figure 6. The power inverter transfer function and the equivalent stator windings of the three-phase IM.

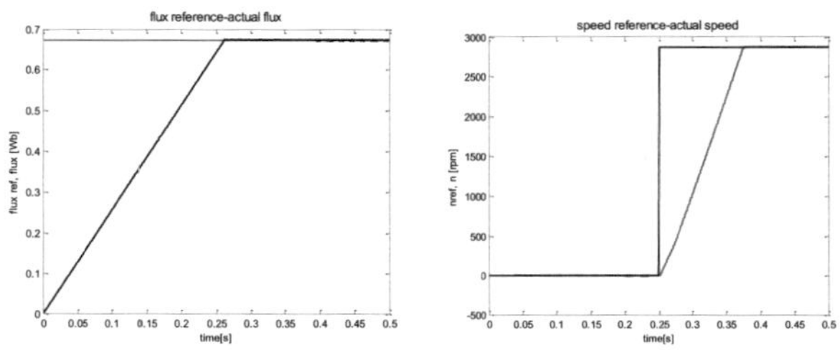

Figure 7. The comparison between the input and output signals of the FOC with IM drive.

criterion. Based on the nameplate motor data presented in **Table 1**, the Matlab script providing automatically the parameters of the tuned speed and current controllers is depicted in Ref. [8].

Parameters	Value	Parameters	Value	Parameters	Value
Rated power	P_n=45 kW	Power factor	0.86	Rated flux	0.67 Wb
Rated speed	2980 rpm	Rated efficiency	0.92	J	0.15 kgm^2
R.m.s line voltage	380 V	R_s	0.0763Ω	I_{mN}	18.86 A
Rated frequency	50Hz	R_r	0.0261Ω	Rated load torque	144 Nm
Rated current	87 A	$L_{s\sigma}$	0.0009 H		

Table 1. The nameplate data of the three-phase induction motor.

3. The modulation techniques

In order to increase both the efficiency and the harmonic contents of the three-phase power inverter, four types of modulation strategies have been implemented in real time using dSpace platform. The implemented modulation strategies [9–12] are as follows: (1) sinusoidal modulation— Sinusoidal Pulse Width Modulation (SPWM), (2) Third harmonic insertion—Third harmonic Pulse Width Modulation (THPWM), (3) space vector modulation—space vector Pulse Width Modulation (SVPWM) and (4) optimized modulation—optimized modulation Pulse Width Modulation (OPWM). The first approach is the *rectangular pulse modulation*. This method raises harmonic distortion problems, and the amplitude of the fundamental component of the output voltage is fixed. However, the frequency could be varied. In order to obtain adjustable amplitude, a derivative method has been deducted: the *quasi-rectangular pulse modulation*. Nowadays, the most used method is the *Sinusoidal Pulse Width Modulation* (SPWM) due to the introduction of three degrees of freedom—the phase, the frequency and the amplitude of the fundamental component of the alternative output voltage. Additionally, the harmonics content of the output signal is considerably diminished. Another issue of the static power inverters is the DC link voltage utilization. By means of the modulation techniques significant improvement of the power inverter efficiency and harmonics content of the output signals could be obtained. The most common modulation technique is the sinusoidal pulse width modulation (SPWM). The SPWM method introduces an important advantage: generates the high order harmonics, therefore the lower weight of the filter inductance is obtained in order to compensate them.

In **Figure 8**, the three-phase power inverter schematic is shown ($v_{s1,n}$, the voltage between the 1 and n potentials; V_{dc}, the DC link voltage).

Based on the analysis of the control signal and the carrier signal, by deducting the conduction time, t_{on}, the analytic formula of the SPWM duty cycle is deducted as follows:

$$d_1 = \left[\frac{1}{2}\right] + \frac{1}{2}\left(\frac{v_{s1,n}}{V_{dc}/2}\right) \tag{5}$$

The modulator has been implemented in real time through the dSpace platform (**Figure 9**):

By using the implemented cascade control (**Figure 2**) in the dSpace platform, the speed reference has been followed and the following three-phase inverter output currents have been obtained (**Figure 10**).

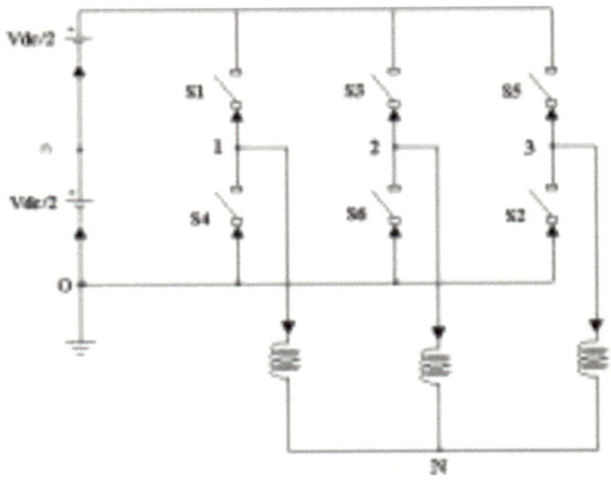

Figure 8. Three-phase power inverter schematic.

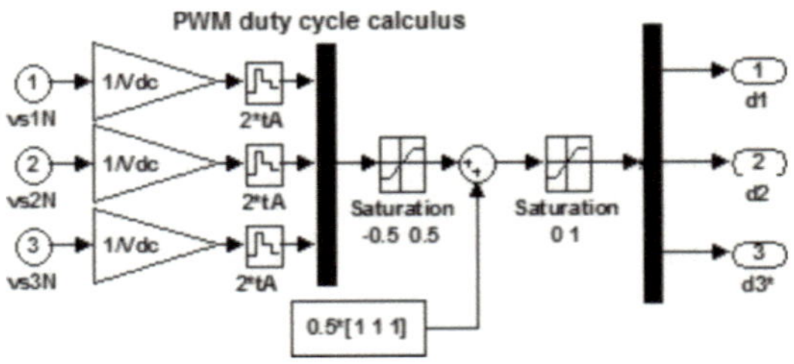

Figure 9. Matlab-Simulink implementation of the sinusoidal PWM modulation technique.

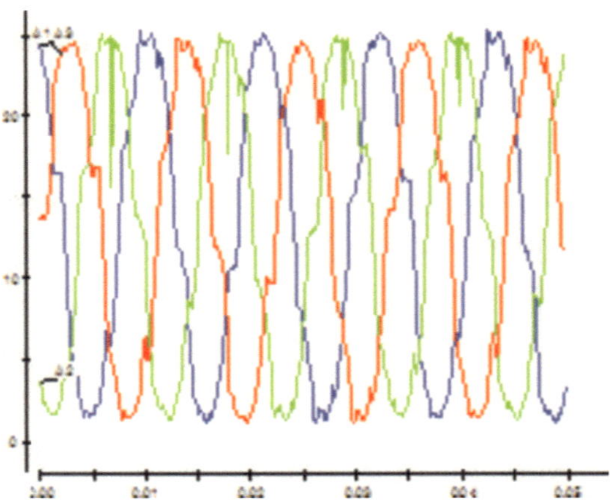

Figure 10. The output currents of the three-phase power inverter with sinusoidal modulation.

The real-time implementation results of the speed control are shown in **Figures 11** and **12**. The speed is reduced through the speed reference. In **Figure 11**, the performances of the speed control are shown (The step signal is the speed reference and thedelayed one is the feedback speed).

http://dx.doi.org/10.5772/intechopen.68217

By using the adequate modulation techniques, the efficiency of the power conversion could be increased.

By inserting the *third harmonic component* in the sinusoidal waveform (**Figure 13**), the three-phase reference voltages are obtained (**Figure 14a**). The improved efficiency and the decreased harmonic content will be obtained by means of the *space vector modulation technique* (**Figure 14b**). In order to minimize the number of the switching, *the optimized modulation technique* could be applied (**Figure 14c**).

The optimized modulation could be applied for the isolated three-phase load as IM. The optimized modulation (**Figure 14c**) minimizes the number of commutations by subtracting a

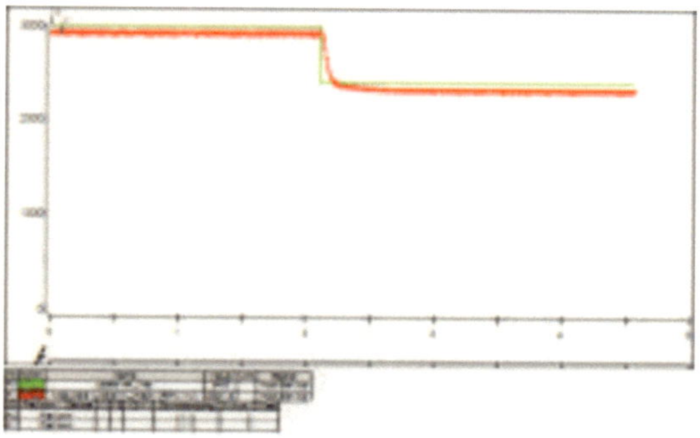

Figure 11. The real-time speed control of the IM based on the dSpace platform.

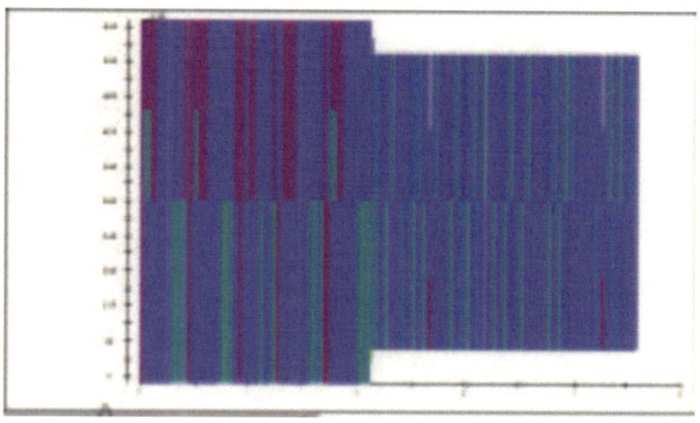

Figure 12. The corresponding three-phase power inverter output voltages.

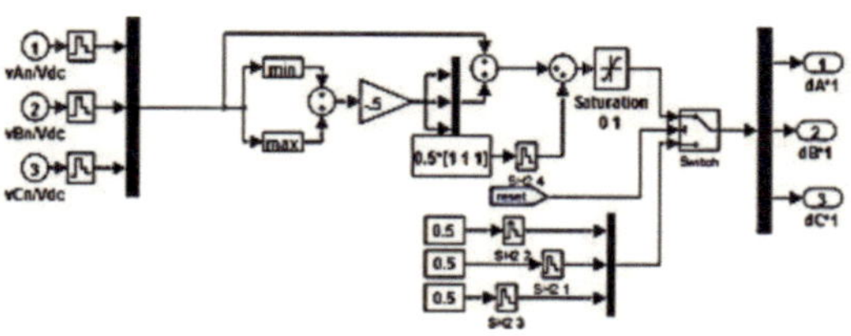

Figure 13. The Simulink implementation of the *third harmonic component* insertion and of the sinusoidal PWM.

zero sequence signal, u_z^* [Eq. (6)], from the sinusoidal voltages [Eq. (7)]. In this way, the usage of the DC link voltage is increased. In **Figure 15**, the dSpace implementation of the OPWM is shown.

$$u_z^* = -\min\left[\frac{U_{dc}}{2} - \max(u_a^*, u_b^*, u_c^*), \quad \frac{U_{dc}}{2} - \min(u_a^*, u_b^*, u_c^*)\right] \tag{6}$$

$$\begin{aligned} u_a^{**}(t) &= u_a^*(t) - u_z^*(t) \\ u_b^{**}(t) &= u_b^*(t) - u_z^*(t) \\ u_c^{**}(t) &= u_c^*(t) - u_z^*(t) \end{aligned} \tag{7}$$

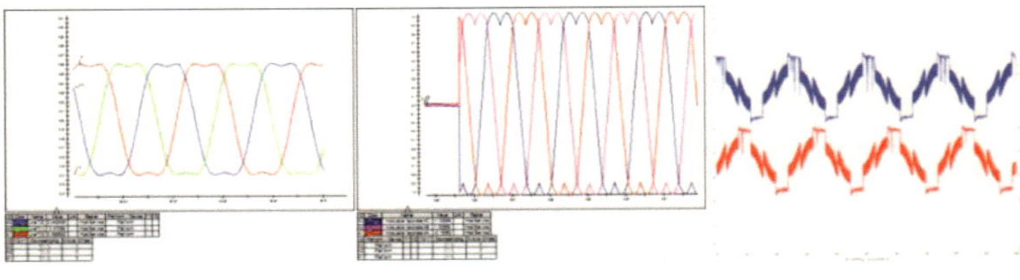

(a) (b) (c)

Figure 14. The reference signals for different modulation techniques (a) the TH-PWM, (b) space vector SV-PWM and (c) optimized modulation (O-PWM).

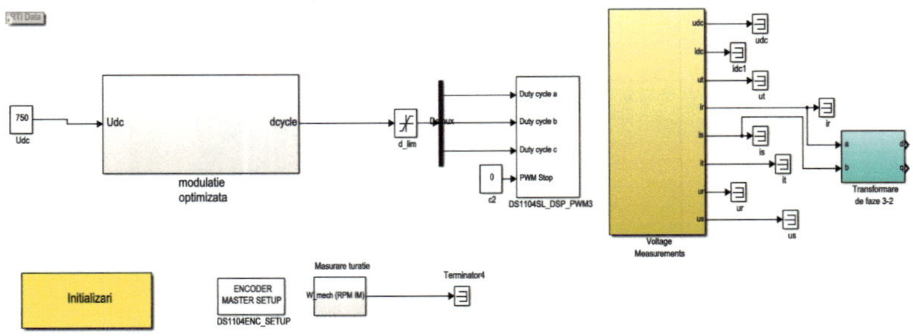

Figure 15. Block diagram of the dSpace implementation of the optimized modulation.

4. The advanced control of the three-phase induction machine based on power inverter

In this section, the advanced control of the IM drive is provided taken into consideration the on-line estimation of the controller parameters. Therefore, the variation of the parameters does not affect the controller performances. The presented adaptive control is robust to unmodelled parameter variations and structural uncertainness. The model reference adaptive control, in direct form, unnormalized of the three-phase induction machine has been used (**Figure 16**)

http://dx.doi.org/10.5772/intechopen.68217

[13, 14]. The adaptive control $u(t)$ contains two components: the gradient ($\theta_g \in \mathcal{R}^{2n}$) and the variable structure ($\theta_v \in \mathcal{R}^{2n}$):

$$u(t) = \theta^T(t)v(t), \theta = \theta_g + \theta_v \tag{8}$$

In **Figure 17**, the Simulink implementation of the rotor field-oriented control of the IM (supposing the constant magnetizing current at the rated value) is shown.

In order to obtain the adaptive control (8), the vector of the filtered signals should be known:

$$v(t) = [v_u^T(t) \quad v_{y_p}^T(t) \quad y_p(t) \quad r(t)]^T \in \Re^{2n_p}, \tag{9}$$

where $r(t)$ is the reference of the adaptive system; y_p (t) the output signal of the process; $v_u \in \Re^{n_p - 1}$ the dynamic of the filter connected at the control, $v_y \in \Re^{n_p - 1}$ is the dynamic of the filter connected at the output of the plant [14]

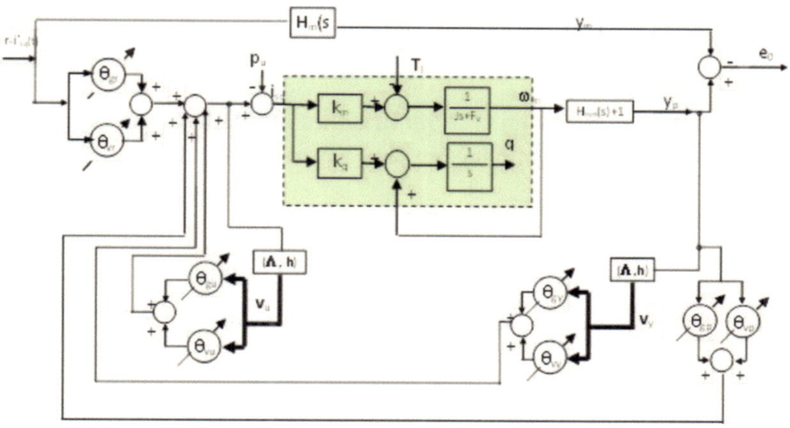

Figure 16. The block diagram of the three-phase IM model reference adaptive control.

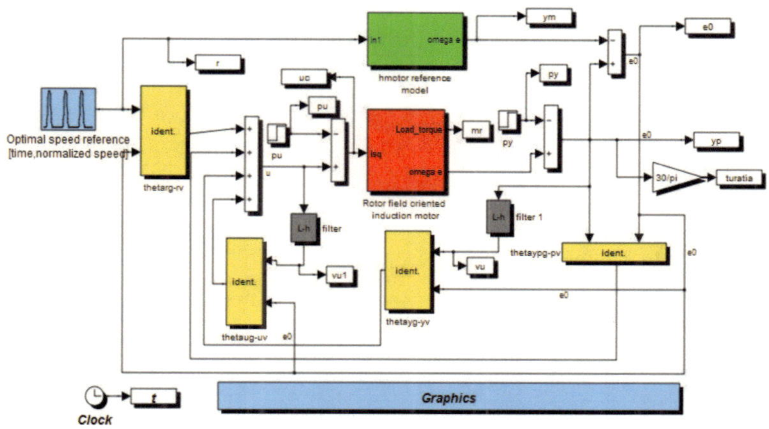

Figure 17. Simulink implementation of the model reference adaptive control with unity relative degree.

$$\begin{cases} \mathring{\mathbf{v}}_{\mathbf{u}} = \Lambda\mathbf{v}_{\mathbf{u}} + \mathbf{h}u \\ \mathring{\mathbf{v}}_{\mathbf{y}} = \Lambda\mathbf{v}_{\mathbf{y}} + \mathbf{h}y_p \end{cases} \tag{10}$$

where

$$\begin{aligned} v_u(s) &= (s\mathbf{I} - \mathbf{\Lambda})^{-1}\mathbf{h}U(s) \\ v_y(s) &= (s\mathbf{I} - \mathbf{\Lambda})^{-1}\mathbf{h}Y_p(s) \end{aligned} \tag{11}$$

The adaptive control is formed by two components (**Figure 18**): gradient and variable structure. The specific parameters of the above-mentioned components are calculated via **Figure 18.**

The parameters vector with gradient-adjustment control [14, 15]:

$$\theta_g \in \Re^{2n_p} = \begin{bmatrix} \theta_{gu}^T(t) & \theta_{gy_p}^T(t) & \theta_{gp}(t) & \theta_{gr}(t) \end{bmatrix}^T \tag{12}$$

The dynamics components of the gradient parameters vector are calculated as (**Figure 18**):

$$\mathring{\theta}_{gu} = -\gamma_g \cdot \text{sign}\,(k_p) \cdot v_u \cdot e_0 \tag{13}$$

The gradient component assures stability and makes smooth transient response and zero tracking error. The asymptotic performances will be assured by gradient-adjustment component.

The variable structure control [14]

$$u_v(t) = \theta_v^T v(t), \tag{14}$$

where the parameters vector with variable structure adjustment control are as follows:

$$\theta_v(t) = \begin{bmatrix} \theta_{vu}^T(t) & \theta_{vy_p}^T(t) & \theta_{y_p}(t) & \theta_r(t) \end{bmatrix}^T. \tag{15}$$

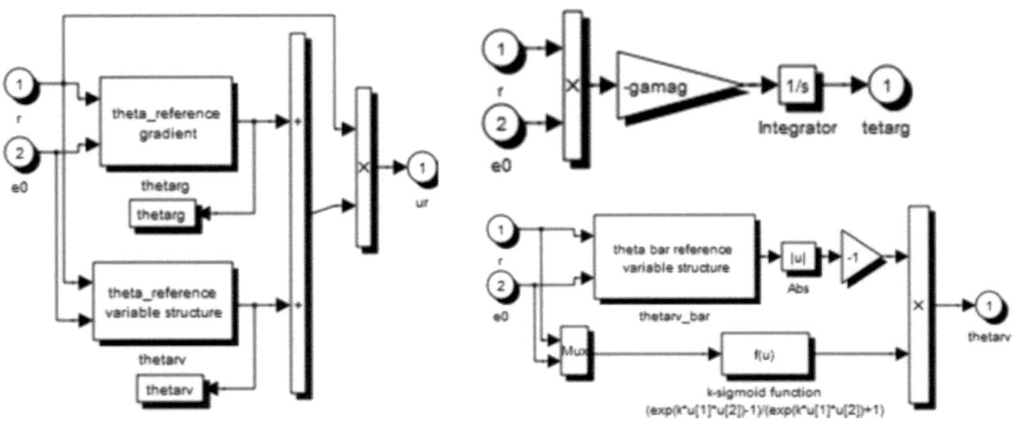

Figure 18. The gradient and variable structure components of the adaptive control. The calculation of the specific parameters (θ_g, θ_v).

http://dx.doi.org/10.5772/intechopen.68217

The variable structure adaptive component assures a fast time response of the control by introducing a *signum* function [14]:

$$\theta_{vu} = \overline{\theta}_{vu}\text{sign}\,(k_p)\;\text{sign}\,(e_0 v_u). \tag{16}$$

In order to eliminate the small oscillations around the equilibrium point, the enhanced feature is included in the variable structure adaptive component [16, 17] through the *k*-sigmoid function [14, 15]:

$$\theta_{vu} \cong \overline{\theta}_{vu}\frac{e^{ke_0\mathbf{v}_u} - 1}{e^{ke_0\mathbf{v}_u} + 1}\text{sign}\,(k_p) \tag{17}$$

where

$$\overset{\circ}{\overline{\theta}}_{vu} = -\boldsymbol{\lambda}_{vu}\overline{\theta}_{vu} - \gamma_v|e_0 v_u|. \tag{18}$$

In **Figure 19**, calculation of the parameter $\overline{\theta}_{vu}$(mentioned in **Figure 18**) is based on Eq. (18):

In **Figure 20**, according to Eq. (10), the filtered vector calculation is shown.

In **Figure 21**, the Simulink implementation of the IM mathematical model is shown.

The differential equation of the circular motion is solved by using the Laplace transform as in **Figure 22**. The Reverse Park and Clarke transformations are implemented in Simulink [13] as in **Figures 23** and **24**. The performances of the model reference adaptive control are shown in **Figures 25–27**.

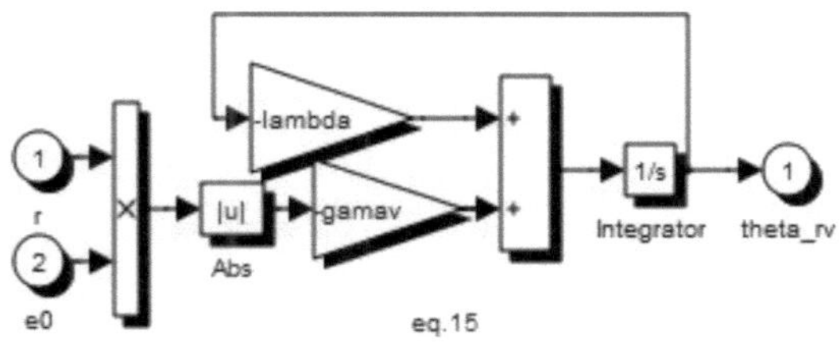

Figure 19. The parameter $\overline{\theta}_{vu}$ calculation.

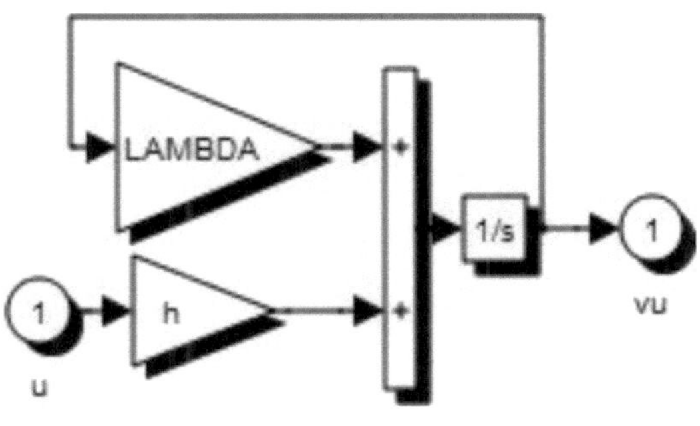

Figure 20. The filtered vector calculation.

The advanced control consists of the model reference adaptive control of three-phase induction machine. It could be noted that a robust MRAC drive system has been provided, the speed being almost insensible to load torque disturbances (**Figures 25–27**). **Figure 25** shows the obtained adaptive control (i_{sq}) at the constant rotor-magnetizing current, the reference and the actual values of the speed references and the evolution of the tracking error. Taking into account the gradient and variable structure parameters (**Figure 26**), the resulted adaptive control, i_{sq}, assures the stability, robustness to load variation (**Figure 27**), smooth transient response of the adaptive controller parameters and zero-tracking error (**Figure 26**). The asymptotic performances are assured by the gradient component. In

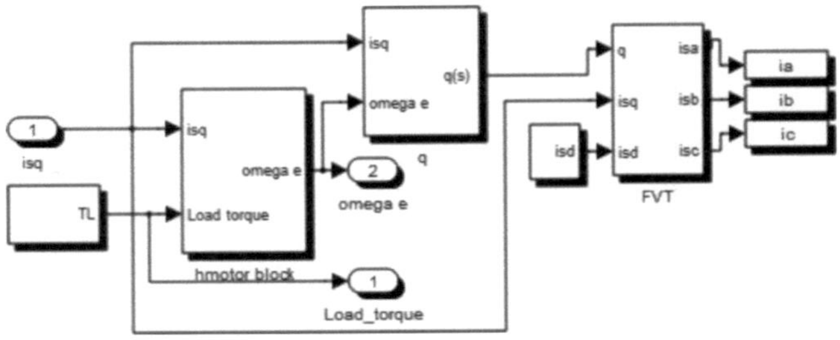

Figure 21. Simulink implementation of the IM mathematical model.

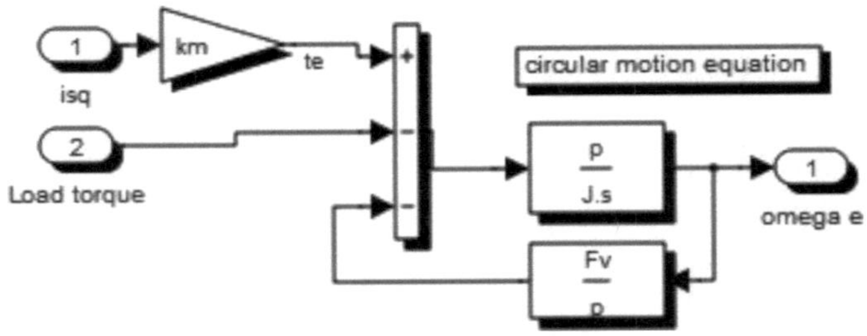

Figure 22. The solution of the mechanical equation.

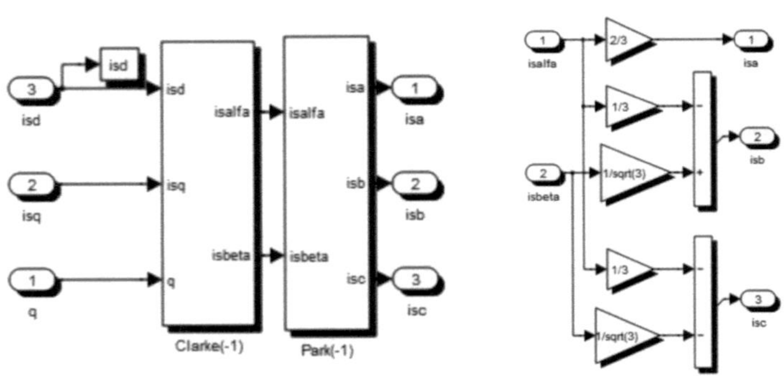

Figure 23. The Reverse Park and Clarke transformations.

http://dx.doi.org/10.5772/intechopen.68217

Figure 27, the three-phase stator currents of the IM under the step load torque variation are shown.

In Annex I, the dSpace implementation of the stator current control of the three-phase IM with sinusoidal modulation is provided.

The first figure from the Annex I contains: (1) at the left side: the reference values of the stator frequency and the r.m.s. reference value of the stator current, the initiation of the control process (pwm_enable), the confirmation of the normal operation of the dSpace platform (dspace_ok); (2) in the middle: the control system; (3) the outputs: the duty cycles, the estimated three-phase stator voltage supply and the measured three-phase stator currents.

The second figure from the Annex I contains the measuring system of the three-phase stator currents, the stator currents references and the control system based on the proportional integral regulators. The outputs of the PI regulators are used in order to generate the sinusoidal modulation duty cycles (duty_abc).

Figure 24. The Simulink implementation of the reverse Park transformation.

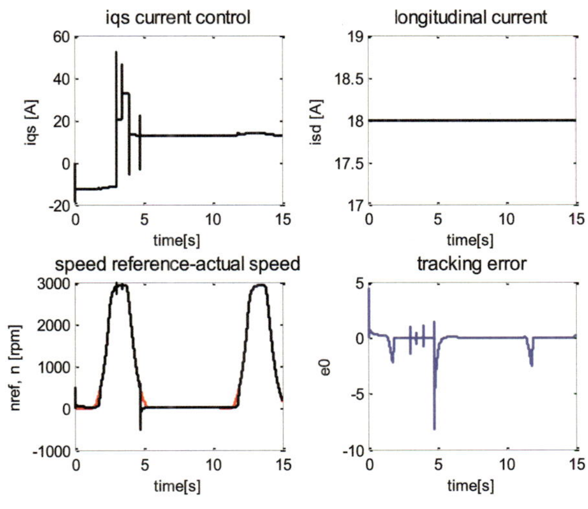

Figure 25. The adaptive control (i_{sq}) under the rated magnetization rotor flux (i_{sd}), the IM drive output (ω, angular speed) and the evolution of the tracking error (e_0).

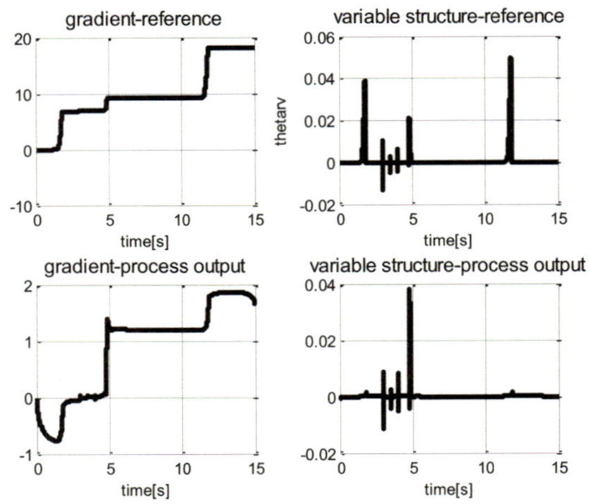

Figure 26. The evolution of the gradient and variable structure parameters.

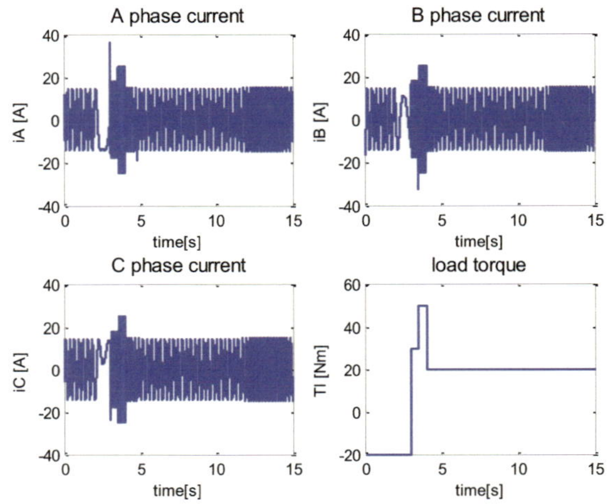

Figure 27. The three-phase stator currents under the step load torque variation.

5. Conclusions

The switching function-based mathematical modelling methodology of the full-bridge single-phase power inverter is provided. The adequate Matlab-Simulink implementation has been shown. The advantage of increasing the switching frequency by two times is taken through the unipolar asymmetric PWM modulation. Additionally, the harmonic spectrum shows a decreasing distortion in spite of the bipolar symmetric PWM modulation [8]. The chapter includes the basic theoretical aspects, followed by mathematical modelling, numerical simulations and implementation of the proposed modulation techniques capable both to increase the DC voltage usage and decrease the harmonic content of the output signals; therefore, by using Matlab/Simulink software, an efficient and clean power converter is obtained. The three-phase power inverter connected to the three-phase induction motor is considered. The four types of the modulation techniques for the three-phase power inverter have been presented and implemented through a real-time platform: sinusoidal PWM, third harmonic insertion PWM, optimized PWM and Space Vector PWM. Moreover, the increased efficiency of the power inverter is obtained through both DC link voltage utilization and harmonic distortion reduction. In order to prove the feasibility of the provided solutions the inductive load based on the three-phase induction machine has been used supplied by means of the three-phase voltage power inverter. The SVM–PWM is considered as optimal switching modulation and due to only one transition between the two switching states excursion takes place. The flat top discontinuous PWM or optimized PWM has the advantages of minimum switching modulation and increased DC link voltage [18] is used [9]. Therefore, higher efficiency is obtained and electromagnetic compatibility is improved [10]. By inserting the third harmonic, the fundamental of the output voltage increases by 15.5% comparative with the sinusoidal PWM. The original advanced electric drive system based on the MRAC has been presented.

Acknowledgements

This work was supported by a grant of the Romanian National Authority for Scientific Research, CNDI–UEFISCDI, project number PN-II-PT-PCCA-2011-3.2-1680.

Nomenclature

ω_{sl}	Slip angular speed
i_{ds} (i_{qs})	Longitudinal stator current component *d* (transversal stator current component *q*)
v^*_{ds} (v^*_{qs})	*d*-axis stator reference voltage (*q*-axis stator reference voltage)
L_s (L_r)	Stator inductance (rotor inductance)
L_m	Mutual inductance
$T_s=L_s/R_s$	Stator time constant
σ	Total leakage factor
T_L	The equivalent load torque reduced to the rotor shaft
$T_r=L_r/R_r$	Rotor time constant
K_T	Torque constant
J	The equivalent inertia moment
p	Number of the pole pairs
R_s, R_r	Stator and rotor phase resistance, respectively
ω_{mR}	Angular speed of the synchronous rotating frame
ω_{sl}	Slip angular frequency
ω_r	Rotor angular speed
$e_{ds}=i^*_{ds} - i_{ds}$, $e_{qs}=i^*_{qs} - i_{qs}$	The *d-q* components of the current error
i^*_{ds}, i^*_{qs}	The *d-q* components of the stator current references
K_p, K_i	The proportional and the integral coefficients

MRAC	Model reference adaptive control
λ_{dr}	The longitudinal (reactive) component of the rotor magnetising flux
λ_{r}	The rated value of the magnetic flux
λ_{qr}	The quadrature (active) component of the rotor magnetising flux

Annex I. dSpace implementation of the stator current loop control

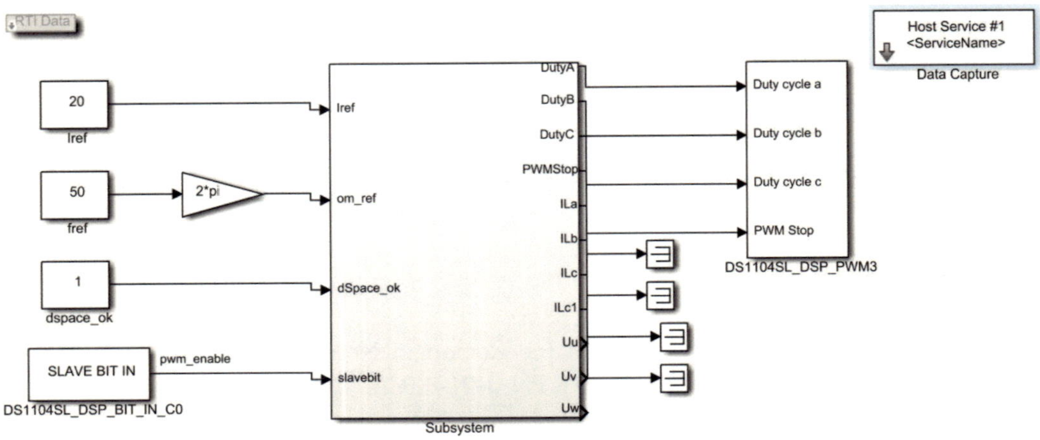

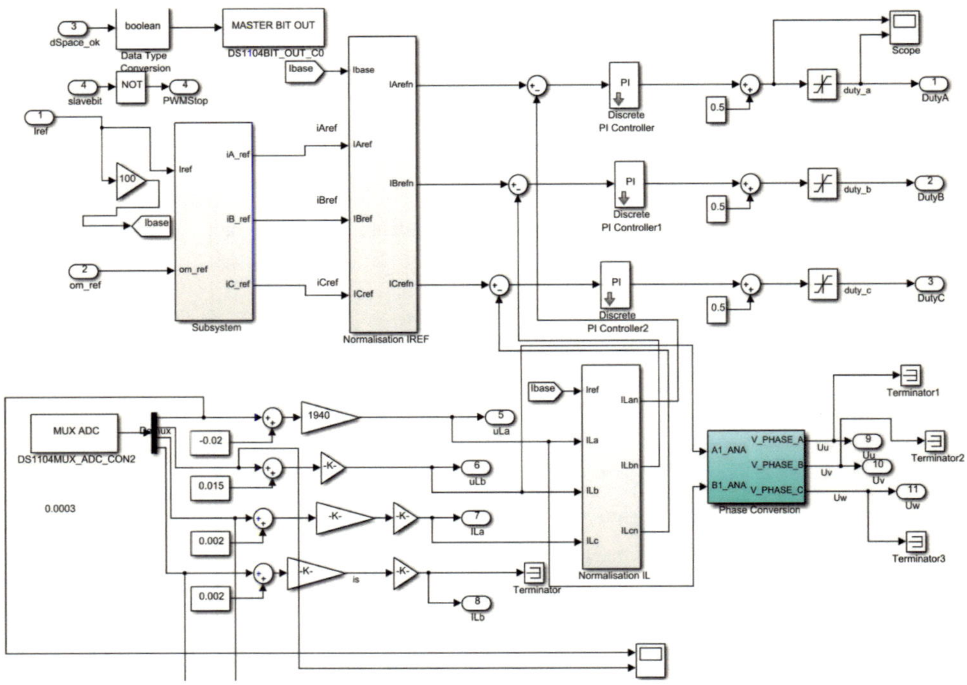

Author details

Marian Gaiceanu

Address all correspondence to: marian.gaiceanu@ieee.org

Integrated Energy Conversion Systems and Advanced Control of Complex Processes Research Center, Dunarea de Jos University of Galati, Romania

References

[1] Keyhani, A., Marwali, M.N., Higuera, L.E., Athalye, G., Baumgartner, G. An integrated virtual learning system for the development of motor drive systems. IEEE Transactions on Power Systems. 2002;**17**(1):1–6.

[2] Mohan, N., Robbins, W.P., Imbertson, P., Undeland, T.M., Panaitescu, R.C., Jain, A.K., Begalke, T. Restructuring of first courses in power electronics and electric drives that integrates digital. IEEE Transactions on Power Electronics. 2003;**18**(1):429–437.

[3] Ayasun, S., Chika, O.N. Induction motor tests using MATLAB/Simulink and their integration into undergraduate electric machinery courses. IEEE Transactions on Education. 2005; 48(1): 37–46.

[4] Saghafinia, A., Ping, H.W., Uddin, M.N., Amindoust, A. Teaching of simulation an adjustable speed drive of induction motor using MATLAB/Simulink in Advanced Electrical Machine Laboratory, 13th International Educational Technology Conference, Procedia—Social and Behavioral Sciences. 2013; 103: 912–921. http://www.sciencedirect.com/science/article/pii/S1877042813038585

[5] Elmas, C., Ali, A.M. Virtual electrical machinery laboratory: A fuzzy logic controller for induction motor drives. International Journal on Engineering Education. 2004; 20(2): 226–233.

[6] Gaiceanu, M. AC-AC Converter system for AC drives. IEE Conference Publication Journal.; British Library, Great Britain: Wrightsons; 2004; 2(498):724–729. DOI: ISSN 0537-9989

[7] Leonhard, W. Control of Electrical Drives. Berlin: Springer-Verlag; 1996.

[8] Gaiceanu, M. Tool of the Complete Optimal Control for Variable Speed Electrical Drives. In: Mr Kelly Bennett, editor. MATLAB Applications for the Practical Engineer. InTech; Croatia, 2014; 339–374. ISBN 978-953-51-1719-3; DOI: 10.5772/57521

[9] Kaźmierkowski, M.P., Krishnan, R., Blaabjerg, F., editors. Control in Power Electronics: Selected Problems. USA: Academic Press; 2002.

[10] Analog Devices Inc. Double Update Mode of PWM Generation Unit of the ADMC401, **AN401**–02. 2000 (January).

[11] Stumpf, P., Jardan R.K., Nagy, I. Comparison of naturally sampled PWM techniques in ultrahigh speed drives. In: International Symposium Industrial Electronics. 2012: 246–251, DOI: 10.1109/ISIE.2012.6237092 Conference: ISIE 2012, At Hangzhou, China, IEEE International Symposium on Industrial Electronics 2012, Article number 6237092, Pages 246-251, 21st. IEEE International Symposium on Industrial Electronics, ISIE 2012; Hangzhou; China; 28 May 2012 through 31 May 2012; Category numberCFP12ISI-ART; Code 91747.

[12] Rathore, A.K., Holtz, J., Boller, T. Synchronous optimal pulsewidth modulation for low-switching-frequency control of medium-voltage multilevel inverters. IEEE Transactions on Industrial Electronics. 2010;57(July):2374–2381.

[13] Gaiceanu, M., Rosu, E., Dache, C., Paduraru, R., Munteanu, T. Adaptive Control of the Three-Phase Squirrel Cage Induction Motor with Load Torque Estimator. Proceedings of the International Conference on Optimisation of Electrical and Electronic Equipment, OPTIM. Brasov, Romania, 2012.

[14] Filipescu, A. A compound law for adaptive and variable structure control with sigmoid function for the process with unitary relative degree model. In: The 9th Symposium on Modelling and Identification systems; SIMSIS8'94; Galati. Romania.1994. pp. 14–18.

[15] Ioannou, P., Fidan, B. Adaptive control tutorial. The Society for Industrial and Applied Mathematics. 2006. 2006/ xvi + 389 pages/ Softcover/ ISBN: 978-0-898716-15-3/ http://bookstore.siam.org/dc11/.

[16] Orłowska-Kowalska T., Korzonek M. Stability analysis of MRASCC speed estimator in motoring and regenerating mode. Power Electronics and Drives 2016 | Vol. 1 (36), No. 2 | pp.113–131.

[17] Kumar, R., Das, S., Syam, P., Chattopadhyay, A.K. Review on model reference adaptive system for sensorless vector control of induction motor drives. IET Electric Power Applications. 2015;9(7):496–511.

[18] Balogh, A., Varga, E., Varjasi I. DC link floating for grid connected PV converters. World Academy of Science, Engineering and Technology, International Journal of Electrical, Computer, Energetic, Electronic and Communication Engineering. 2008;2(4):629–634.